Sikorsky
H-34

Sikorsky

H-34

An Illustrated History

Lennart Lundh

Schiffer Military/Aviation History
Atglen, PA

ACKNOWLEDGMENTS

Historians are hostage to the distance between Then and Now. Across that distance, measuring almost fifty years, much material about the H-34 has been lost. This book is an attempt to decipher the remaining records, which are often confused as well as confusing. Despite the help of many people, all errors in sifting through the pieces of the puzzle are mine alone. Anyone with corrections or fresh information is encouraged to contact me through the publisher.

John Hairell was kind enough to read and comment on a draft of this book; his observations are beyond price. Sid Nanson shared the fruits of his many years of research on Sikorsky helicopters, and nudged me to reconsider "common knowledge" on the subject. Responding to my requests for crew recollections and observations were Don Alberts, Jackson Arney, Gerald Bickford, David Bjork, Mac Echols, Roger Ek, Paul Gregoire, Pat Kenny, Al Kerst, Ed Kozak, Leonard Martinez, Max Mitchell, Bob Steinbrunn, and Luther Stephens; thank you, gentlemen, for bringing the H-34 to life.

My gratitude is extended to the following individuals: Rick Albright, Joe Baugher, Dana Bell, Tsahi Ben-Ami, John Benton, Rudy Binnemans, Bob Burns, Regina Burns, Hubert Cance, Dick Casterlin, Ken Conboy, Steve Darke, Sergio dos Santos, Jean-Paul Dubois, James Fore, Fred Freeman, Jack Friell, Ned Gilliand, Albert Grandolini, Dan Hagedorn, Dave Hansen, Rollin Hatfield, Leif Hellström, Bob Hoppers, Pete Hughes, Larry Jacks, John Kerr, Tim Kerr, Dennis Kuykendall, Hank Lapa, Terry Love, Elof Lundh, Eduardo Luzardo, John MacGregor, William McGlaugn, Ben Marselis, Dave Menard, Jim Mesko, Rob Mignard, David Moore, Wayne Mutza, Bob Nokes, Bill Norton, Jorge Nuñez Padin, Gogor Nurharyoko, Mike O'Rourke, Kirsten Tedesco, Kevin Pace, Bill Parrish, Dick Phillips, Don Raczon, Byron Reynolds, Skip Robinson, Geoff Russell, Frank Shelly, Hidenori Shinmura, L.B. Sides, Ken Snyder, Norm Taylor, Johan Vanderwei, Stefan Vanhastel, Jeroen van Reijmersdal, Gustavo Wetsch, Kai Willadson, Nick Williams, Leon Wohlert, Carolyn Wright, Mike York, and Benjamin Yu.

Thanks are extended to the following organizations: Aircraft Metals, Components, Engines and Parts; Air Force Historical Research Agency; American Aviation Historical Society; Canadian Forces Photo Unit; Candid Aero-Files; Central Florida Aircraft Photographs (CFAP); Coast Guard Historian's Office; Defense Reutilization and Marketing Service; Defense Technical Information Center; the Dwight D. Eisenhower and John F. Kennedy Libraries; Federal Aviation Administration; Los Angeles County Sheriff's Department Aero Bureau; Military Aircraft Photographs (MAP); Moore Aviation; Museo Nacional de Aeronáutica de Chile; National Aeronautics and Space Administration; National Archives and Records Administration; Emil Buehler Library, National Museum of Naval Aviation; National Technical Information Service; Piasecki Aircraft; Service Historique de l'Armée de l'Air (SHAA); Sikorsky Aircraft Corporation; Small Air Force Clearing House; Tucson Air Museum Foundation of Pima County; Turbomeca; UT/Dallas Special Collections; Westland Helicopters; Winnebago Industries; Air One, Aris, Construction, Hi-Lift, Northwestern, and Orlando Helicopters; Brazilian, Italian, and Japanese Navies; the National Air and Space, US Air Force, Militaire Luchtvart, Third Cavalry, and Israeli Air Force Museums; US Army Aviation Museum, Transportation Museum, and Troop Command.

Book Design by Ian Robertson.

Printed in China.
ISBN: 0-7643-0522-0

We are interested in hearing from authors with book ideas on related topics.

Published by Schiffer Publishing Ltd.
4880 Lower Valley Road
Atglen, PA 19310
Phone: (610) 593-1777
FAX: (610) 593-2002
E-mail: schifferbk@aol.com
Please write for a free catalog.
This book may be purchased from the publisher.
Please include $3.95 postage.
Try your bookstore first.

CONTENTS

Title Page: **HSS-1Ns of HS-7 pose for the camera. First is 147990 (c/n 58-1230), second is 148004 (c/n 58-1256). (Photo KN-7188 courtesy of National Archives via Rob Mignard)**

DEDICATION

For Lin, with love, and Jake Dangle, with thanks

In memory of
Cpl. Thomas Edward Anderson, USMC
Lt. Gerald Charles Griffin, USN
Sgt. Richard Elmer Hamilton, USMC
HM2 Gerald Owen Norton, USN
Sgt. Wayne Jerald Pendell, USMC
1Lt. Michael Joseph Tunney, USMCR
LCpl. Miguel Angel Valentin, Jr., USMC
KIA 6 October 1962, South Vietnam

Chapter 1
Nuts and Bolts

Sikorsky's model S-58, used in armed forces around the world, almost didn't come into military service. Bell's HSL-1 won the contest for a Navy anti-submarine warfare (ASW) helicopter. In Army and Air Force competitions, it was bested by the tandem-rotor Piasecki H-21. The Marine Corps wanted Sikorsky's model S-56, the heavy-lifter known in the military as the HR2S-1 and H-37. Without military models to drive interest, and experience to bring about improvements, it is unlikely there would have been commercial models.

Circumstances conspired to reverse these ill fortunes. The HSL-1 was unsuitable for its shipboard ASW role, leading the Navy back to the HSS-1 Seabat. With Piasecki's facilities taxed by H-21 production for the Air Force, the Army placed orders for the H-34A Choctaw. This ran against the expressed better judgment of Air Force test and evaluation teams. Late in the H-34's American service life, the Air Force used ex-Navy H-34s for search and rescue. When development problems plagued the H-37 project, Navy and Army experience with the Seabat and Choctaw convinced the Marines of the need for ever larger numbers of the HUS-1 Seahorse.

As a result, over the years since 1954, military and civil versions of the S-58 have served the armed forces, governments, and commercial enterprises of over forty nations. In all, Sikorsky produced 1,825 (including the four prototypes and one duplicated construction number), and Sud Est in France built 135 under license. Appendix One details Sikorsky's production by model and year. Only about one H-34 in ten still exists, and perhaps half of those are airworthy. Still, it is likely that there will be examples flying, at least in civilian service, to celebrate the type's fiftieth anniversary in the year 2004.

Designations

With widespread and lengthy service, as well as the 1962 introduction of the American Tri-Service Designation system, the H-34 has come to carry a bewildering assortment of des-

The final XHSS-1, 134670 (c/n 58-4x), on the Sikorsky ramp at Bridgeport, CT. For publicity purposes, "ARMY" was added to the photograph. (Photo A0643 courtesy of US Army Transportation Museum)

Naval Reserve HSS-1 138490 (c/n 58-48) on the ramp at NAS Glenview, IL, on 29 April 1961. (Photo by P. Stevens courtesy of Norm Taylor)

Backed by fog and rain, all-weather HSS-1N 147984 (c/n 58-1177) as a testbed for the Naval Weapons Systems Command sometime between late 1960 and late 1962. This was the first of the v-leg HSS-1Ns. (Photo courtesy of MSgt David W. Menard, USAF, Ret.)

ignations. Because of this, it is common to find photos and descriptions mislabeled as to the particular model being viewed. There have even been two unofficial designations (the H-34B and OH-34D) which never came into service but persist in the literature, as well as a DH-34D found briefly in the US Civil Aircraft Register (although there is no evidence for the use of H-34s as drones). Appendix Two gives the known designations for military aircraft.

Common Design Features

In addition to being known by a variety of names, the H-34 was, as the comments in the chart of designations suggest, actually a variety of helicopters. All models, however, shared a number of features. Dimensional and performance specifications are provided in Appendix Three.

The airframe was an all-metal, semi-monocoque structure. The major sections were a forward engine compartment, a cockpit above and behind the engine, a transmission deck behind the cockpit, a main cabin box below the transmission deck and behind the engine, and a low-mounted tailcone and pylon. It was skinned largely with magnesium. The reverse-tricycle fixed landing gear included a fully swiveling tail wheel. Power for the main and tail rotors, both of which consisted of four all-metal blades, was provided by the nose-mounted Wright R1820-84 Cyclone engine. The engine was cooled by screened intakes around the nose and a fan mounted behind the engine. The main rotor was driven by a transmission connected to the Cyclone by a shaft running from the clutch compartment through the forward cabin bulkhead, the forward cabin, and the main transmission deck. Shafts and gearboxes

H-34A 53-4497 (c/n 58-55) lifting a Double-Wasp engine. (Photo S17087 courtesy of Sikorsky Aircraft Corporation)

With full French-pattern nose armor, a Sud Est-built H-34A lifts one of its Sikorsky counterparts in Algeria. (Photo B87-840 courtesy of SHAA via Albert Grandolini)

CH-34C 55-4498 (c/n 58-540), 15th Medical Detachment, 7th Army, at Bad-Kreusnach, Germany on 11 August 1966. (Photo by Brink courtesy of Norm Taylor)

Finishing its service days with Marine Corps Reserve HMM-772, UH-34D 148777 (c/n 58-1332) went on to work with Midwest Helicopter Airways as N90561. (Photo by R. Esposito courtesy of MSgt David W. Menard, USAF, Ret.)

In partially subdued markings, late-production UH-34D "Yankee Victor Six" from H&MS-15 flies over the California coast. (Photo courtesy of Paul Gregoire)

Minus its amphibious gear, ex-Marine UH-34E 145727 (c/n 58-885) of HC-4 waits out the rain at NAS Lakehurst, NJ, in May of 1966. (Photo by R. Esposito courtesy of Norm Taylor)

HPS-Helicopter Leasing GmbH's S-58C D-HAUG (c/n 58-836) shows some of the type's unique external features. This aircraft, the last S-58C built, served earlier with Chicago Helicopter, Sabena, and the Belgian Air Force. (Photo courtesy of Meravo)

Briles Wing and Helicopter's S-58ET N33602 (c/n 58-727) was once a West German H-34G.I. (Photo courtesy of CFAP)

Extensively reglazed, and sporting radar above the engine, British Airways Helicopters' S-58JT G-BCLN (c/n 58-1539, ex-H-34G.III 150753) runs up with the assistance of a ground cart. (Photo courtesy of CFAP)

Engine clamshell doors were not made of "sterner stuff," as evidenced by the damage to SH-34J 148030 (c/n 58-1298) before museum restoration in Kissimmee, FL. (Photo 950720-8 courtesy of the author)

Seen earlier with HMM-772, N90561 displays the "high-stack" flame-dampening exhaust and its characteristic clamshell cutout, along with the air intake screens. Aircraft converted from the original to the later exhaust often had modified clamshells with both openings. (Photo 080080-19 courtesy of the author)

The original "low-stack" exhaust configuration exited through a cutout in the bottom of the port clamshell. The aircraft is SH-34J 143954 (c/n 58-754), awaiting restoration at Hi-Lift Helicopters in Florida. (Photo 950731-14 courtesy of the author)

running internally along the spine of the tail boom and rotor pylon provided power for the tail rotor.

Access to either side of the cockpit was typically gained by way of kicksteps on the forward fuselage, with the large, sliding side windows providing room for entry. It was also possible, if cramped and awkward, to pass between the cockpit and main cabin. This was done by raising the bottom pan of the pilot's or co-pilot's seat and crawling through the resulting gap between the rear cockpit bulkhead and floor. Entry to the main cabin was easier, with a large, aft-sliding door provided on the starboard side. This presented a three-foot sill, however, which might have been ideal for loading and unloading cargo but provided a definite challenge for combat-laden troops. To correct this, various step configurations were developed, both in the factory and the field. Along with these avenues of entry and exit, the windows mounted in

Wright Cyclone access doors from the front of ex-HH34J 48019 (c/n 58-1278). Additional cooling intake screens are above the landing light. (Photo 950720-35 courtesy of the author)

From the right, the revised nose for the P&W PT6T-3 and -6 turbines. Although it looks substantially longer, the "T-nose" only added six inches to the length of the aircraft. S-58ET N580US (c/n 58-1673) from Midwest Helicopter was built as H-34G.III 150804. (Photo 921111-1 courtesy of the author)

The port nose and twin exhausts of Midwest Helicopter's S-58ET N4247V (c/n 58-1547, ex-H-34G.III 150754). (Photo 890524-86 courtesy of the author)

either side of the cabin fuselage could be knocked out in case of emergency.

Also common was a folding tail rotor pylon, developed to meet Navy requirements for compact shipboard stowage, which could be swung through 180 degrees to the left and rested against the main boom. In addition, the main rotor blades were designed to fold along the fuselage, further decreasing space requirements.

Design Differences

As experience was gained, and as new roles were filled, changes were required. Thus, if the basic package was common across models, neither it nor the contents were identical. This was true even within a given model, where changes didn't justify a new designation. The differences discussed here apply to Sikorsky-built, American-employed aircraft unless noted. Most probably also apply to foreign examples, but manuals for non-American forces, giving exact details of modifications (especially for radios, avionics, electronics, and instrumentation), have not been available.

Nose Doors

The Wright Cyclone was covered by vertically split clamshell doors on all models, with variations in the port door for different exhaust configurations. Hinge dampers and open-door latches were fitted during the production cycle. Clamshell doors are also employed on the TwinPac-powered conversions, albeit of a different design.

Power Plant

For all military models, the Wright R1820-84 Cyclone series was standard. Some were also built under contract by Lycoming. The commercial and export equivalent was the Wright 989C9HE-2. The original military models were the -84A (Army) and -84B (Navy and Marines). With the advent of ASE, an engine governor and fluid coupling impeller drive

Behind the clamshell doors, mounted so that most work could be accomplished without stands or ladders, was the Wright R1820-84-series Cyclone engine. (Photo courtesy of Robert N. Steinbrunn)

A 6x6 wrecker helps pull the engine of a Mississippi National Guard CH-34C in February, 1972. The prop-shaft end of the engine faces up in this shot, with the housing for the cooling fan just above the cylinders. (Photo courtesy of Mississippi National Guard)

With clamshells and engine removed, ex-Uruguayan CH-34C A-066 (c/n 58-887) displays the firewall, cooling fan cutout, and open rear access hatch. To the right and level with the top of the access hatch is the shelf for the inverter. (Photo courtesy of the author's collection)

were added, heavy-duty pistons were incorporated, and the engines became the -84C and -84D, respectively. Engine reworks were accomplished for all services by the Naval Air Rework Facility at NAS Jacksonville. Piston powerplants were mounted at a thirty-four degree forward angle with the propeller shaft facing aft. The S-58T conversion utilized Pratt & Whitney (Canada) PT6T-3 or PT6T-6 TwinPac turbines behind a modified nose. A single HSS-1F, powered by a pair of General Electric T-58 turbines, was tested by Sikorsky and the Navy. In France, two aircraft were experimentally fitted with twin Turbomeca Bastan turbines. Neither installation went into production, although both were critical to future powerplant developments.

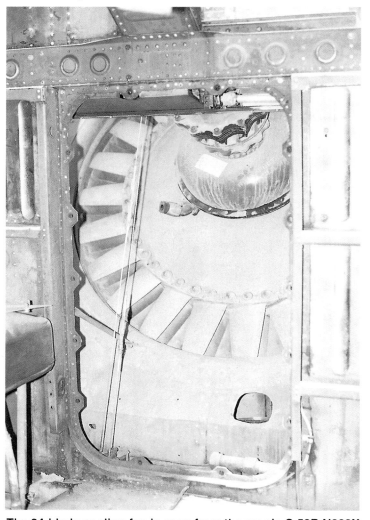

The 24-blade cooling fan is seen from the rear in S-58B N882X (c/n 58-403). (Photo 890715-3 courtesy of the author)

Doors open, Los Angeles County Sheriff's S-58T N392JK (c/n 58-740, ex-H-34G.I) undergoes routine maintenance at Long Beach, CA. (Photo 930821-10 courtesy of the author)

The low-stack and mid-mounted single-barrel exhausts created vision problems for pilots during night operations because of exhaust flames. The Sikorsky response was the high-stack FDE (Flame-Dampening Exhaust) shown on 143954. (Photo 950731-13 courtesy of the author)

Oil Tanks
On the H-34A and HUS-1, two bladder cells with manual oil dilution carried 10.5 gallons. On the HUS-1A and HSS-1, the cells had a 12.4 gallon capacity. On all commercial models, a single stainless steel hopper with automatic oil dilution carried 10.4 gallons.

Engine Controls
The original rod-and-link controls were replaced with cable on the H-34A beginning in November, 1956, and on the HSS-1 from January, 1957 (BuNos 141591, 141602 and subsequent). Cable was standard on all other military models, and on all commercial models. In all cases, the carburetor heat control remained rod-and-link.

Firewall Area
This was titanium on all military models, stainless steel on all commercial models.

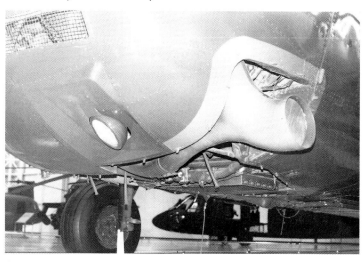

Cyclone exhausts came in three basic styles. The low-stack single exhaust extended out from the bottom left of the nose, as on CH-34A 53-4526 (c/n 58-88), displayed at Fort Rucker. (Photo courtesy of the author)

Because the low-stack exhaust presented a fire hazard on H-34s fitted with floats, an interim measure was a single-barrel style placed just above the mid-line of the nose. Here without its floats, "XM-17" was HUS-1A 145725 (c/n 58-860), assigned to HMX-1. (Photo 67-312-29 courtesy of Dwight D. Eisenhower Library)

PT6T-3 exhausts on S-58T N64CH (c/n 58-787, ex-UH-34D 145738) of the LA County Sheriff's Department Aero Bureau. (Photo courtesy of the author)

H-34A 56-4332 (c/n 58-747) crashed due to engine failure in the mountains north of Asiago, Italy, on 29 April 1958. Its misfortune provides a good view of the exposed engine and oil cooler areas that required armor plating in combat. (Photo courtesy of Gerald Bickford)

The armor on VNAF UH-34G 141598 (c/n 58-256) has been moved back to ease maintenance. (Photo courtesy of Donald Raczon)

A late-production UH-34D shows the extent of the protection provided by the Marine-pattern armor used in Vietnam. (Photo courtesy of Donald Raczon)

Armor

The exposed oil cooler, thinly covered engine components, and high crew seats were weak points when the H-34 came under fire. The French added armor protection in Algeria, but the Marines were forced to relearn this lesson from hard experience in Vietnam. In US service, armor was only fitted to the Seahorse. It was also found on various models in Southeast Asian services. The Army did a separate evaluation process, but armor was not installed on Choctaws since they did not see combat in Vietnam. The French-designed armor was of a different pattern, and was also used on Israeli examples.

Winterization

Winterization equipment for pre-heating the engine, transmission, and cockpit was tested on one H-34A, and available but not factory-fitted on all H-34As. It was available as loose equipment on the HUS-1 variants as well, except that HUS-1Ls were delivered with engine, transmission, cabin, and cockpit heaters fitted. Similar equipment was not available for the ASW or commercial models as delivered.

The original kicksteps had a springed-hinge cover. The steps on commercial models, such as Chicago Helicopter S-58C N876 (c/n 58-310), were located just enough farther forward than on military aircraft to cause pilots difficulty. (Photo courtesy of the author)

HH-34F 1343 (c/n 58-1068) shows the steps with the covers removed, a normal practice in service. (Photo 950625-37 courtesy of the author)

Battery

On the S-58C, a thirty-six volt battery was standard, with a second, twenty-four volt battery optional. On all other models except the H-34A (twenty-four volt only), the twenty-four volt battery was standard and the thirty-six volt optional.

Auxiliary Power Unit

The APU, which could be accessed by ground carts through an external receptacle, was originally located in the tail cone, aft of the radio compartment. The APU and external receptacle were moved just behind the engine compartment beginning with c/n 58-702.

Generator

On all commercial models, a three-hundred-amp, engine-driven generator was installed. All military models had the generator mounted on the transmission. For the H-34A, it was two-hundred amps. The HUS-1G and HSS-1N had three-hundred-amp units mounted on both the transmission and engine. All other military models had a four-hundred-amp generator.

Inverters

The H-34A and all commercial models had two 250-amp inverters mounted forward of the firewall. In the same location,

1343 is displayed at the Florida Military Aircraft Museum with an early flat-paned sliding cockpit window. The plate at the corner of the canopy is a night-flying glare shield. (Photo 950625-47 courtesy of the author)

N90561 carries a bulged sliding window, which provided the pilots with greater visibility than the earlier style. (Photo 080080-22 courtesy of the author)

The later style kicksteps were easier to use and less of a maintenance headache. UH-34D 150227 (c/n 58-1585) is displayed at NAS Pensacola as a LH-34D. (Photo 950820-25 courtesy of the author)

most other models had a 250-amp and a 1,500-amp inverter. The HUS-1G, HUS-1L, and HSS-1N had 1,500-amp units. Additionally, the HUS-1G and HSS-1N carried a one-hundred-amp unit in the aft compartment.

Auxiliary Servo System
On military models, a 1,500-psi reservoir was fitted forward of the firewall. On commercial models, a 1,000-psi reservoir was mounted on the transmission deck.

Main Blades
These were identical on all models. However, a heavier spar was introduced on the H-34A and commercial models in November, 1956, and on Navy and Marine models in January, 1957.

The instrumentation on the co-pilot's side of early H-34As was minimal, as seen in this 1958 shot. (Photo courtesy of Gerald Bickford)

The cockpit of H-34G.III 150741 (c/n 58-1513) holds yet another instrument and control pattern. (Photo 34990B courtesy of Sikorsky Aircraft Corporation)

UH-34D 148768 (c/n 58-1319) provides one example of later instrumentation. (Photo 870717-5 courtesy of the author)

Kicksteps

Rectangular steps, each with a hinged cover, were placed on both sides of the forward fuselage on all models through c/n 58-1393 for UH-34Ds and 58-1428 for SH-34Js. They were replaced in production by open-faced, semi-circular steps. The latter were easier to negotiate, and overcame maintenance problems inherent in the hinged covers of the earlier style. The steps were located farther forward on commercial models than on military models.

Cockpit Windows

A single sliding window was fitted level with the crew seats on either side, and used for entrance and exit. These were originally flat-paned, and later bulged as an aid to visibility. The five-panel windscreen was equipped with wipers and a

Radio controls, which varied from model to model, and the throttle sat between the pilot and co-pilot, as in this shot of an Army H-34A from 1958. (Photo courtesy of Gerald Bickford)

The transmission deck on S-58 N504 (c/n 58-895, ex-CH-34C 57-1717). Commercial aircraft without transmission deck covers are a common sight. (Photo 900915-6 courtesy of the author)

The overhead switch panel, circuit breaker and fuse panel, rotor brake, and lights were mounted between the cockpit seatbacks. A data case is above the co-pilot's seat. The aircraft is 150741. (Photo 34990-E courtesy of Sikorsky Aircraft Corporation)

forced-air defroster. Five overhead panels provided upward visibility. Blind-flying panels could be installed internally for the side and front windows.

Flight Controls
The HSS-1N, CH-34C, and later HUS-1 variants were equipped with auto-stabilization equipment. Cyclic and control stick details differed across and within models.

Instruments
All military models had instrument panels with similarity of function but variations in grouping. On the H-34A prior to s/n 54-3000, only the dual tachometer, manifold pressure gage, and airspeed indicator were repeated on the co-pilot's side. Comparable pilot's and co-pilot's groupings were the norm. The standard commercial instrument panel provided for optional gyro and co-pilot units in addition to the pilot's basic grouping.

Cabin Windows
On the S-58C, there were six on the port side (two of which were escape-type) and four on the starboard. On other commercial models and all military models, there were two on the port side (escape type) and one on the starboard. Bulged observer's windows were common on HUS-1A and HUS-1G aircraft. Extensive re-glazing is common on ex-military aircraft in commercial service.

The main rotor head, with two blades removed, on N882X. (Photo 910421-6 courtesy of the author)

N882X with the transmission cooling vents in place. (Photo 890715-5 courtesy of the author)

Maintenance personnel of the 22nd Special Warfare Aviation Detachment make use of the cockpit roof and port workstand in mid-1962. The object of their labors is the first H-34A, 53-4475 (c/n 58-15), after conversion to H-34C standards. (Photo PN7395 courtesy of US Army Aviation Museum)

Cabin Doors
On the S-58C, there were two hinged passenger doors on the starboard side. On other commercial models and all military models, one removable, sliding cargo door with a single window was fitted on the starboard side. An enlarged, tinted window was fitted in the cargo door of VIP models. Modifications to the cargo door are common on ex-military aircraft in commercial service.

Cabin Step
The most common step configuration on standard military models was a single steel-tube rung fitted below and at the front edge of the cargo door. In the eighteen-seat troop configuration, the seats across the cargo door folded out and down to form steps. On the S-58C, a double platform step was fitted beneath each passenger door. On other commercial models, a single steel tubing step was fitted below and the width of the cargo door. Military VIP models had a single set of steps similar to those on the S-58C.

Cabin Floor
An aluminum structure on all models, the cabin floor was reinforced on the S-58C and had provisions for a sixth fuel cell on the forward tank of other commercial models. The HSS-1 was fitted with a sonar well. All other military models were fitted with skid rails, tiedown rings, and troop seat/litter rack fittings. Army Choctaws had a Plexiglas window to aid in sling operations, begining in 1960.

Forward Fuel Tanks
The HSS-1 and HUS-1G were fitted with a 115-gallon, five-cell bladder. Other military models were equipped with a one-

For aircraft with the starboard rescue hoist, the transmission cover/workstand was fitted with a springed-hinge hatch to allow the workstand to fold down around the hoist, as on 1343. (Photo 950625-31 courtesy of the author)

hundred gallon, five-cell, self-sealing bladder. All commercial models had a six-cell, 123-gallon bladder.

Center Fuel Tanks
Standard on all military models was a 70.7-gallon, three-cell bladder. All commercial models were equipped with a seventy- gallon, three-cell bladder.

Aft Fuel Tanks
All military models carried a 92.3-gallon, three-cell bladder. A ninety-two-gallon, three-cell bladder was optional on commercial models.

Troop Seats
The HUS-1 provided canvas seating for twelve troops (plus an observer in the HUS-1A), the H-34A for twelve or (in production from s/n 56-4313, with retrofitting beginning in

143954 shows the removable saddle and braces necessary when the main rotor blades are folded. (Photo 950731-18 courtesy of the author)

Since missions were often flown with the cabin windows removed, a wind guard was fitted forward of the port windows on aircraft such as 1343. (Photo 950625-45 courtesy of the author)

LH-34D 145717 (c/n 804) carries both bubble and flat cabin windows. (Photo 911020-8 courtesy of the author)

Reglazing, such as on N90561, is common on ex-military aircraft. (Photo 080080-12 courtesy of the author)

For the convenience of passengers, S-58Cs like N876 were heavily glazed and had a door on each of the two compartments. (Photo courtesy of the author)

The standard military cabin door on 1343 has a center window. Some aircraft had the window on the aft edge of the door. (Photo 950625-20 courtesy of the author)

S-58Cs were equipped with a baggage compartment. On military models, this position was occupied by the crew chief's seat. (Photo courtesy of the author)

1960) eighteen. The HUS-1G could be equipped with seating for ten troops along with the standard radio operator and observer bucket seats. HSS-1s had bucket seats for the sonar operator and his relief. Commercial models other than the S-58C offered twelve-man, canvas seating as an option.

Passenger Seats

In the S-58C, there was seating for six passengers in each of the two compartments into which the cabin was divided. Army and Marine (and, it is assumed, Navy) VIP H-34 variants provided bench seats for ten (three three-seat units and a single-seat unit). In addition to the five known HUS-1Zs, plush interiors were also manufactured for all twenty-two HUS-1As and four HUS-1s (BuNos 143971-143974). The S-58C passenger seats were upholstered to airline standards and customer requirements. In the VIP H-34s, the finish was simpler than

VH-34C 56-4320 (c/n 58-718) at the Army Aviation Museum has the panoramic cabin door window typical of the VIP models. (Photo courtesy of the author)

Cabin doors were usually reglazed along with the fuselage on ex-military examples like N90561. (Photo 080080-8 courtesy of the author)

Doors were sometimes also heavily modified, such as this "screen porch" change on N64CH. (Photo courtesy of the author)

in the S-58C but far better than the troop seats of other military models.

Litters
Fittings were installed for mounting eight litters in all models except the S-58C, which had none, and the HUS-1G, which had a rack for a single Stokes basket stretcher.

Soundproofing
Cockpit and full-cabin soundproofing were standard on all models except the H-34A and HUS-1; these had soundproofing only for the cockpit and transmission tunnel.

Because the door tracks were open-ended, a stop was fitted at the rear-most position. (Photo 950625-20 courtesy of the author)

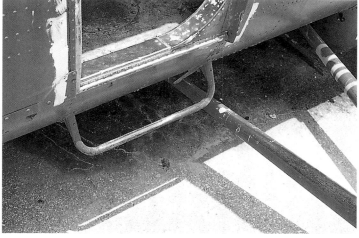

A small step at the forward edge of the cabin floor lip was standard on Seabats such as 148030. (Photo 950720-10 courtesy of the author)

Awaiting its fate in Indiana long after retirement, Sud Est-built SA-168 displays the full-width step commonly found on 12-seat H-34As. (Photo 070782-30 courtesy of the author)

VH-34C 57-1725 (c/n 58-910), on display at the Army Transportation Museum, has typical two-riser VIP steps. (Photo 950903-3 courtesy of the author)

Baggage Compartment
On the S-58C, a baggage compartment with hinged exterior door was placed on the starboard side between stations 82.5 and 112. There was no special compartment on military or other commercial models.

Radio Compartment
On all military and commercial models other than the S-58C, the radio compartment was aft of the cabin, between stations 246 and 296, and partitioned by a removable, zipper-closure curtain. On the S-58C, it was located above the baggage compartment.

RIGHT: When 18-troop seating was installed, the seats across the cabin door folded out to provide a step. CH-34C 56-4331 (c/n 58-739) was with the Idaho National Guard when this photo was taken. (Photo courtesy of Rollin Hatfield)

S-58C steps were similar to those on VH-34s, but were permanently mounted. (Photo courtesy of the author)

Heating
Heaters rated at 50,000 BTUs were standard equipment on the S-58C, H-34A, and HUS-1 (which could be equipped with 200,000 BTU kits). Other HUS-1 variants were equipped as ordered, the HSS-1s with 25,000 BTU factory installations, and the remaining commercial models with 100,000 BTU installations. The tail boom space immediately behind the radio compartment housed the heater, and ducts ran forward, along the upper walls of the cabin, into the cockpit.

Hand-held Fire Extinguisher
One liquid fire extinguisher was stored in the H-34A cabin, with CO_2 being used in Navy and Marine Corps models. All commercial models had a cockpit fire extinguisher, and a hand-held unit was standard in both cabins of the S-58C.

Like other VH-34s, 57-1725 also has a large handrail forward of the cabin door. (Photo 950903-2 courtesy of the author)

Fuel cells were located under the cabin floor. The forward cells have been removed from S-58 N51803 (c/n 58-1280, ex-SH-34J 148021), which awaits scrapping or restoration. (Photo 950720-32 courtesy of the author)

Overhead, the tunnel for N882X's tail rotor drive shaft leads away from the floor of the transmission compartment. (Photo 890715-18 courtesy of the author)

143954's radio compartment shelves have been stripped of equipment. The heater compartment and tail cone are aft of the rear bulkhead. The warning sign reads, "Caution No Personnel No Stowage Aft." (Photo 950731-7 courtesy of the author)

Seabats had side-by-side cabin seating for the sonar operator and his relief. Undergoing restoration, 148030's cabin also holds an accumulation of spare or removed parts. (Photo 950720-18 courtesy of the author)

Israeli Air Force S-58B 12 (c/n 58-691) fitted with full military soundproofing. (Photo S23793B courtesy of Sikorsky Aircraft Corporation)

Some Seabats carried an internal auxiliary fuel tank. The filler pipe is just above the tail wheel. Inverter and battery access plates are to the left. 143954 is in civilian hands here as N72RH. (Photo 950731-5 courtesy of the author)

The sonar cable reel, motor, and electronics were stored inside the cabinet that occupies the forward section of the cabin. This French Navy-operated, Sikorsky-built HSS-1 is equipped with an internal auxiliary tank, just visible on the right. Part of the starboard weapon rack may be seen in the lower left. (Photo S22734 courtesy of Sikorsky Aircraft Corporation)

Fire Detection
The commercial models had fire sensors in the engine and heater compartments. Of the military models, only the H-34A and VIP variants had a sensor, which was located in the engine compartment.

Automatic Fire Extinguisher
Automatic fire suppression was provided only in the VIP variants of the military models. The S-58C had equipment in both the engine and heater compartments, while other commercial models had it only in the engine area.

Pyrotechnics
All Navy and Marine Corps models had provisions for a flare gun and cartridges, as well as a floatable smoke canister. No similar provisions were made in H-34As. Tail-mounted parachute flares were optional on commercial models.

Bottom Structure
Originally magnesium, this became aluminum on the HSS-1 from c/n 58-21 and on the H-34A from January, 1957. All HUS-1 variants had an aluminum structure, as did all commercial models except the S-58C (which had an aluminum structure with stainless steel keel beam). Fuel dump pipes were run along the fuselage bottom under protective covers.

Cargo Sling
A cargo sling was optional on commercial models and was not standard on the HSS-1, but was retrofitted to UH-34Gs and UH-34Js. The original Manning, Maxwell and More four-thousand pound capacity hook and sling was replaced in 1956

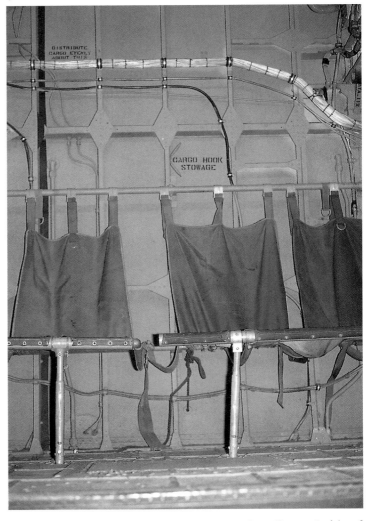

Part of the 12-man troop seat arrangement on the port side of 148768 at the National Air and Space Museum. The seats could be folded against the cabin wall to allow use of litters or the cabin floor. (Photo 870717-1 courtesy of the author)

Litters mounted in 150741. (Photo S28453B courtesy of Sikorsky Aircraft Corporation)

Looking forward in the cabin of 148768. (Photo 870717-2 courtesy of the author)

with an Eastern Rotorcraft A-60 five-thousand pound rig. This coincided with the strengthened structure of the later aluminum fuselage tubs. Army reports show that the change was welcomed, but not totally unopposed. The 506th Transportation Company, when asked for comments by the Aviation Test Board, responded, "The position of this organization relative to any method or device to increase the allowable payload of the H-34 without increasing power output or re-stressing the dynamic components can be stated briefly: we feel that we have a superior 1 1/2 ton helicopter; that it would be a waste of money and a detriment to the program to convert this superior 1 1/2 ton helicopter to a mediocre, or marginal, 2 1/2 ton helicopter."

Tail Pylon
Designed as a space-saving measure for crowded carrier decks, the tail pylon was hinged to fold 180 degrees to port on all models but the S-58C. On some S-58Ts, this capability was removed to save weight.

Main Gear
On all H-34As, HSS-1s prior to BuNo 147984 (c/n 58-1177), HUS-1s prior to BuNo 148053 (c/n 58-1165), and all commercial models, the main gear consisted of a single vertical oleo and single bent horizontal strut (commonly referred to as "bent-leg") on either side of the forward fuselage. This configuration was prone to structural failure, and was the cause of ground resonance accidents. Subsequently, a single vertical oleo and two straight horizontal struts commonly called "v-leg") were employed. Although it has been estimated that modifying existing aircraft to the latter style gear would require two thousand hours of labor per aircraft, there is photographic evidence that this was done in at least two cases (LH-34D 145717, c/n 58-804, and RCAF H-34A 9632, c/n 58-202).

The radio operator's seat in 1343 is mounted farther back than the sonar operator's in a Seabat. Part of the rescue basket shows in the foreground. (Photo 950625-7 courtesy of the author)

The HH-34F radio operator's console. (Photo 950625-1 courtesy of the author)

While not luxurious, the furnishings in VH-34C 56-4320 are an improvement over canvas seats. Curtains were standard on the VIP aircraft. (Photo 940920-9 courtesy of the author)

Emergency Floats and Amphibious Gear
In American service, doughnut floats for all three wheels were standard only on the HUS-1A and VIP models. Pop-out floats mounted on the hub of each main wheel were common on foreign examples, and remain common on ex-military models in commercial service. In this installation, there is a permanently inflated neoprene bag fitted within the tailcone. Hot dog-style floats were optional (and unusual) for commercial models other than the S-58C, which was not delivered with a flotation system.

Rescue Hoist
A six-hundred-pound capacity hoist was standard on all military models, and optional on all commercial models except the S-58C. On H-34As, it was originally installed only on every fifth helicopter. The hydraulic pump was mounted on the transmission deck. Controls were located above the cabin

door and on the control sticks. Installation of the hoist required a redesigned starboard-side engine workstand, with a hinged panel to allow passage of the hoist.

Auxiliary Fuel Tanks
The HUS-1 variants could be fitted with an external 150-gallon auxiliary drop tank, fitted on the port side of the aircraft. Experimentally, the Army fitted various tanks to either side of an H-34 in April of 1961; they were pleased neither with the improved range nor with the way the installation blocked the cabin door. Some HSS-1s, distinguished by a higher-mounted forward fuel filler cap, could carry a twenty-nine-gallon auxiliary tank at the forward end of the cabin.

Navigation Lights
Port and starboard: All models had a single, flashing red light to port and a green light to starboard. These were mounted

VIP soundproofing included filling the gap between the bulkhead at station 85 and the cockpit. (Photo 940920-8 courtesy of the author)

The folding tail pylon was latched on the starboard side. (Photo 911020-21 courtesy of the author)

The S-58C cabin was split into two compartments. Detailing such as the soundproofing was more to airline standards than the military VIP ships. The aircraft is D-HAUG, pictured earlier. (Photo courtesy of Meravo)

The forward passenger compartment in D-HAUG shows off the leather panels and carpeted floor of the S-58C. (Photo courtesy of Meravo)

flush with the fuselage. With the introduction of night-flying capabilities, they caused reflections in the cockpit and vertigo for the crew. To solve this problem, the lights were mounted on extensions referred to as "elephant ears."

Canopy tub: On military models only, there was a flashing white light.

Tail: H-34A, flashing yellow; HUS-1 variants, flashing white; HSS-1, none; commercial, red and white.

Beacon Lights
On all models, a rotating, flashing, high-intensity red light was mounted on the tip of the tail pylon. A smaller light was used on commercial models than on military aircraft.

Landing Lights
On all models, a 450-watt, fully swiveling, retractable light was mounted in the port engine clamshell door. On commercial models, a fixed, six-hundred-watt light was mounted in the starboard clamshell. Optional on commercial models was a six-hundred-watt light at the rear of the cabin belly.

Passing Lights
An optional fifty-watt red lamp in the starboard clamshell door was available on the S-58C only.

Radios and Avionics
Common radio and avionics suites are listed below. As the list indicates, installations varied within models during production. There were also variations for specific missions.

Seabat:
AIC-4A interphone/radio control
APA-89 video coder (HSS-1N)
APN-22 or -117 (HSS-1N from BuNo 147984) radar altimeter
APN-97 or -97A radar navigation system (HSS-1N)
APX-6 IFF radar id set (HSS-1N)
AQS-4C, -4D or -5 (HSS-1N from BuNo 147984) sonar

Only folding the tail pylon along the port side of the fuselage made the large H-34 a practical fit for carrier stowage. 148030 demonstrates the reduction in length. (Photo 950720-23 courtesy of the author)

With the pylon folded, the tail rotor disconnect coupling becomes visible on 143954. The data plates on the pylon and tail cone are useful, if not foolproof, ways to identify H-34s. (Photo 950731-8 courtesy of the author)

With the top of the tail pylon removed, the drive shaft-tail rotor coupling can be seen. The aircraft is H-34J N87717 (c/n 58-1269, ex-SH-34J 148011) of the **LA County Sheriff's Department.** (Photo courtesy of the author)

The early style gear was prone to structural failure and inducing ground resonance. Seen from the rear on N882X, the source of the nickname "bent-leg" is apparent. (Photo 890715-14 courtesy of the author)

ARA-25 UHF direction finder
ARC-2, -2A, or -39 (HSS-1N from BuNo 147984) HF transmitter/receiver
ARC-27A (prior to BuNo 143864) or -55 UHF transmitter/receiver
ARN-21 UHF radio navigation
ARN-30A Omni
ADF-14B (prior to BuNo 143864), ARN-41A, or -59 (HSS-1N from BuNo 147984) LF radio compass
Choctaw:
AIC-12 interphone/radio control
ARA-31 FM homing (from s/n 56-4284)
ARC-6 or -44 FM liaison transmitter/receiver
ARC-12 or -73 VHF transmitter/receiver
ARC-55 UHF transmitter/receiver (from s/n 56-4284)
APN-32 or ARN-12 marker beacon (from s/n 57-1684)
ARN-30A Omni (from s/n 57-1684)

The main purpose of the tail pylon, on 145717 or any other H-34, was to support the tail rotor. The rotor met the transmission under the bulbous cover. (Photo 911020-2 courtesy of the author)

As a rule of thumb, aircraft below c/n 58-1165 did not have the later style gear. One estimate puts the conversion effort for such a change at 2,000 hours of labor. 145717 is one of the exceptions. The mount for the early gear strut is clearly visible above the forward strut of the "v-leg" gear. (Photo 911020-14 courtesy of the author)

The starboard main wheel and tire on 53-4526. (Photo courtesy of the author)

145717's tail gear, from the port side, with the strut compressed. (Photo 911020-5 courtesy of the author)

The strut mounts and brakes on the port wheel of N64CH. (Photo courtesy of the author)

With strut extended, N90561's tail gear makes the aircraft sit higher. (Photo 080080-14 courtesy of the author)

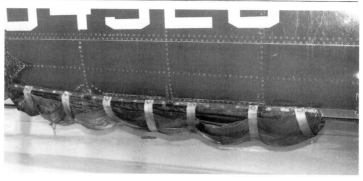

When stowed, as on 56-4320, the rear floats lay along the belly of the tail cone. (Photo courtesy of the author)

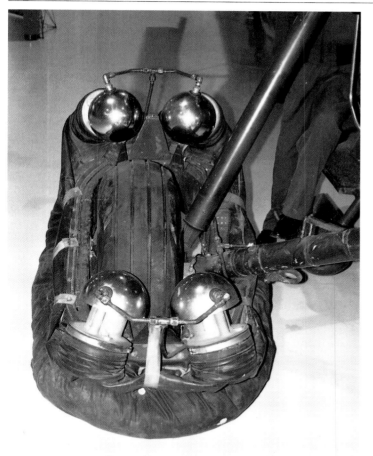

HUS-1A and VH-34 doughnut floats surrounded the main gear. The silver-topped units on 56-4320 are the compressed-air inflation system. (Photo courtesy of the author)

A hub-mounted canister float and its air bottles on v-legged N87717. (Photo courtesy of the author)

A HUS-1A of HMX-1 shows the size of the inflated doughnut floats, and how high they kept the aircraft above the water. The later canister floats, which looked like giant balloons when deployed, were designed to make the aircraft salvageable but not amphibious. (Photo courtesy of Sikorsky Aircraft Corporation)

This Belgian Air Force HSS-1 has a single air bottle. The float has been removed for maintenance. (Photo courtesy of Rudy Binnemans)

The standard 600-pound rescue hoist was mounted on the starboard side, just above the cabin door. It appeared on most military aircraft, including 150227. (Photo 950820-29 courtesy of the author)

There was nothing simple, as the Army complained during tests, about the mount for the 150-gallon external fuel tank, seen here on 145717. (Photo 911020-10 courtesy of the author)

S-58C N867 has both the right-door landing light, unique to the passenger version, and the left-door light standard on all S-58s and H-34s. (Photo courtesy of the author)

As displayed on 1343, the landing light was hinged at the top. It was controlled from the cockpit. (Photo 950625-8 courtesy of the author)

As 145717 shows, the plumbing was every bit as complicated as the tank mount. (Photo 911020-13 courtesy of the author)

ARN-59 LF radio compass (from s/n 57-1726)
Seahorse:
AIC-4A interphone/radio control
APA-89 video coder (HUS-1A)
APN-22 radar altimeter (HUS-1A)
APN-97 radar navigation set (HUS-1A)
APX-6 IFF radar id set (HUS-1A)
ARA-25 UHF direction finder
ARC-2 or -2A (HUS-1 prior to BuNo 147147), -39 HF transmitter/receiver
ARC-44 FM liaison transmitter/receiver
ARC-55 UHF transmitter/receiver
ARN-21 UHF radio navigation
ARN-30A Omni
ARN-41A (HUS-1 prior to BuNo 147147) or -59 LF radio compass
Commercial:
As required by individual customers.

Weapons
Only the Seabat was factory-equipped to carry weapons. A removable, variable-angle launcher for two weapons (one port, one starboard), mounted on a shaft running through the fuselage, was standard on ASW aircraft. The weapons could be salvoed electronically or manually. The shaft and its fittings were modified starting with BuNo 141602 (c/n 58-444); these were internal changes, and are not visible from outside the aircraft. The weapons racks were often removed when ASW duties were not being performed.

The Army, Marines, and French services experimented with a variety of external weapons configurations; only the Marine Corps' TK-1 weapons kit saw trials in Vietnam. Marine units in Vietnam employed field-designed cabin mounts for defensive machine guns. The French Air Force used a mix of externally mounted rockets and fixed machine guns in Algeria, along with a wide range of cabin-mounted machine guns and cannons.

Originally, the port and starboard navigation lights were mounted flush with the fuselage. When the bubble canopy windows were introduced, this became dangerous due to reflections in the cockpit from the flashing lights. (Photo 950731-21 courtesy of the author)

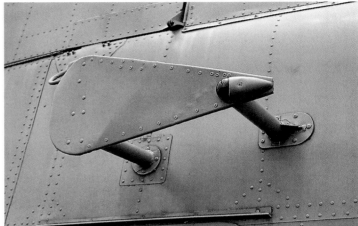

LEFT: Spotlights were often added to the main gear struts, sometimes in pairs facing opposite directions. 150227 has such a paired mount. (Photo 950820-37 courtesy of the author)

The "elephant ear" was designed to shield the cockpit crew from the flashing navigation lights. This feature is sometimes mistakenly identified as part of the antenna suite. (Photo 950625-43 courtesy of the author)

Chapter 2
Crew Notes

I have been fortunate enough to have fourteen H-34/S-58 pilots and crew chiefs share their recollections of missions and insights into the workings of what one calls the DC-3 of helicopters. I am deeply indebted to them for their help. Here, in rough chronological order, are those memories and descriptions.

A.M. Echols III, United States Marine Corps:

I had just transitioned from fixed-wing, propeller-driven aircraft in 1958 upon return from overseas, and reported to Marine Air Group Twenty-six, Second Marine Air Wing, at MCAS New River at approximately the time the MAG was getting the first "giant" HUS-1s to replace their HRS (H-19) helicopters. Castro's coming to power in Cuba heralded my assignment to HMR(L)-262 in October, 1958. I remember having to leave my wife and two boys while visiting my folks in Pennsylvania during Christmas that year. The orders were to report directly to Naval Station Norfolk, and board USS BOXER (CVS-21). She was in the process of transitioning from a WWII fixed-wing carrier to become the Navy's first helicopter carrier, LPH-4, effective 29 January 1959.

I remember my carrier qualification (CQ) as an HUS pilot very clearly. En route from Norfolk to Cuba, BOXER parked off Onslow Beach for twelve hours on 12 January 1959 while HMR(L)-261, HMR(L)-262, and a heavy lift detachment from HMM-461 boarded the carrier. Having boarded at Norfolk, I had an advantage in selecting squadron berthing and getting to know the ship before the rest of my squadron mates. I ensured that we got the most desirable ready room aboard ship.

Evidently the Captain had strict orders, as he embarked for Cuba even before the last of the copters came aboard with their maintenance supplies. We were told that we all would qualify en route, and that the ship would not turn into the wind for carrier qualifications. It was interesting to note that because the ship had a light tail wind as we proceeded south, flight operations commenced on 14 January 1959 and normal CQ procedures were revised to insure flights landed into the wind even though the carrier did not turn. All pilots qualified by establishing the landing pattern to approach over the bow of the ship heading north as the carrier steamed south. If you know anything about normal carrier qualifications, you can imagine how we all felt flying this oddball pattern. The visual cues absolutely had to be ignored, and concentration was doubly reinforced on the LZ to avoid getting

vertigo. The entire air complement eventually CQd without major incident, and we all learned that there was more that one way to skin the cat with our whirlybirds.

Not many people know that it was the Marine Corps, and specifically Sub Unit 1, HMR(L)-262 (with its yellow insignia on the engine clamshell door and the bold "YT" numbering) that pioneered the early capsule recovery techniques for Project Mercury. Foremost, perhaps, was young First Lieutenant Wayne Koons, the indomitable force behind all of the inventive solutions dealing with the helicopter's external hook system, and the so-called "Shepherd's Crook". Wayne, with his engineering background and genius, kept the program on course despite many electromechanical problems (and lack of higher-headquarters support—at one point, I was OIC of the unit, which demonstrated how much emphasis our immediate superiors placed on the project). It was Wayne who took the hook to his home workshop and installed a prototype mechanical override of the electrically actuated hook release system. Reinstalled in the original Project Mercury UH-34D, the override system prevented the hook from accidental releases caused by saltwater vapor shorting the electric system, and provided the co-pilot and crew chief with a mechanical override to release the hook when needed. Using dummy capsules, we tested tensile strength of the cable that attached to the capsule from the helicopter via the Shepherd's Crook, working with the capsule flotation bags on land and in various sea states (I think up to seven-foot swells and even in surf chops). To carry out the mission, the H-34 was tested to beyond normal lift limits, and had to prove its capability for the project's capsule recovery every step of the way.

I was the test pilot on many of these occasions, and proved that the original cable tensile strength was insufficient via an unusual aircraft incident. The dummy capsule was adjacent to and northeast of the runways, and our initial flights were a racetrack pattern paralleling the duty runway. I remember that, because of the prevailing wind, our flight path at one point crossed the intercoastal waterway. I was the unfortunate pilot flying the test mission when an early test cable snapped. The helicopter lurched, I lost all radio contact, and watched as the capsule plummeted into the intercoastal waterway a short distance from a southbound barge. The barge ran aground (mud) to avoid the capsule. In the weeks of paperwork that followed the incident, I had to explain why the cable snapped during normal flight and normal

G forces. I had to complete an aircraft accident (incident) report as the remaining cable attached to the hook snapped into and through the skin of the bird and into the radio compartment, destroying both the UHF and FM components in the compartment. I was lucky the cable didn't snap up into the rotor blades. I even had to file an "Obstacle to Ship Navigation" report until the dummy capsule was removed (without damage) from the waterway. The tug was able to remove the barge from the mud at high tide, but there were other endless reports that related to the incident.

The Ground portion of the Marine Air/Ground team has long tried to get a closer hold of the air assets of the Marine Corps. In my experience as an Air Liaison Officer for the Ninth Marine Regiment at Camp Fuji, Japan, and Camp Hansen, Okinawa, in 1958, and as the Air Officer for Task Force Hotel on LZ Vandergrift (aka LZ Stud) on the DMZ in Vietnam, the Ground Commanders have a great love of having "on-site" control of their very own helicopters. It was this desire that led to the birth of the well deck choppers in the late 1950s and early 1960s, long before the launching of the first carrier built for the helicopter, the USS IWO JIMA (LPH-2).

The UH-34D helicopter was the first helicopter that was envisioned to be part of the Navy's Amphibious Fleet. This was carried out by converting the well deck of selected Landing Ships Dock (LSDs) and the larger type Landing Ships Transport (LSTs). Using this concept, the Navy/Marine Corps converted the portion of the well deck normally used for receiving and launching all types of landingcraft, by installing a $250,000 waterproof curtain separating the covered portion of the well deck from the open area. This "hangar" was thus large enough to house eight H-34s to protect them from the salt air; the magnesium skin of the H-34 would literally melt (corrode) away if not immediately washed in a saltfree solution. This area was also used for aircraft maintenance. Also added to the bow over the well deck was a landing platform suitable for the recovery and launch of the H-34. Finally, a Primary Flight (Pri-Fli) shack was added above the well deck to control flight operations and also direct the hooking of the helicopter to the crane for lowering it to and retrieving it from the well deck.

My first introduction to this type operation was on a Mediterranean cruise aboard the USS DONNER (LSD-20), the second LSD to be refitted for this type operation. 1959 was the maiden voyage for both the ship thus outfitted and our detachment from HMR(L)-262. Major "Smiling Jack" Cosby was the OIC of the detachment. Also assigned in this detachment were Major (later Colonel) Floyd K. Fulton, Captain (later Major General) Greg Corliss, Captain (later Colonel) Bill Beeler, Captain (later Chief of Staff for Senator John Glenn) Phillip Upschulte, and Captain (later GAO Management Evaluator) Mac Echols, among various and sundry officers and enlisted personnel. My duties were flying and working in Pri-Fli.

As part of the Amphibious Fleet, DONNER and USS YORK COUNTY (LST-1175) housed the air crews as well as a regiment from Camp Lejeune. We hoisted, spread, and launched our helicopters one at a time, providing a section of aircraft every eight minutes with a mission to land aboard YORK COUNTY, pick up assault troops, and orbit overhead until the prescribed amount of aircraft were available to carry out the preplanned mission, whether it be day or night operations. The copters were also used in other missions, such as very successful anti-submarine patrol. We operated throughout the Mediterranean, and ashore in Spain, France, Rhodes, Greece, Libya, Morocco, and other places for a host of military and goodwill missions. It was on the island of Rhodes that I met Gregory Peck, who was filming *The Guns of Navarone*, and also where I had a passenger flip his lid and try to jump out of the copter during a night instrument flight.

Jackson Arney, United States Coast Guard:

On 26 September 1960, an Air Force B-47 on evening departure from McDill AFB had the left outboard engine explode at about 1,500 feet. This engine's mount came partially loose and the engine came up against the left aileron, deflecting it upward and putting the B-47 into a series of aileron rolls until the plane struck Tampa Bay inside the Tampa Bay Bridge. All three crewmen ejected and came down in the Bay with only minor injuries. I was a witness to the accident.

An H-34, CG 1343, flown by Lieutenant John Redfield, immediately departed the Coast Guard Air Station at Albert Whitten Field (downtown St. Petersburg). John arrived at the accident site and started to hoist survivors. I was at my home near the Air Station when I saw the B-47 crash, and I went to the Air Station as I knew they would be very busy. When I arrived, I took over radio communications with the helicopter and a Coast Guard forty-foot patrol boat that was underway to the crash.

John said, "I have one." Two or three minutes later, he said, "I have another one," and two or three minutes after that, "I have the third one—we're going in." I lost communications with the helicopter at that time. The patrol boat called that they could not see the helicopter's lights, and believed it had gone in the water.

In the meantime, the Air Station Executive Officer, Commander Don Reed, had arrived. He asked if the other helicopter, CG 1333, was out of check. When told the check was completed but the helicopter had not been test flown, he turned to me and said, "Let's go." We gave the aircraft a thorough preflight and departed for the crash site, which was only six or seven miles from the Air Station.

It was full dark now. We located the crash with lots of floating debris and found CG 1343 with only three or four feet of the tail out of the water. We started a search for survivors. We sighted some reflective tape, and were in a hover trying to determine if it was someone with a flight helmet on when we started recirculating our rotor spray through the rotors and got the "milkbowl" effect. We increased altitude a little, moved away, and it cleared up. We continued the search for several minutes. I had my head out the window when Commander

Reed said, "We are losing rpm. We're going in." I looked at the instruments, and our rpm was down to 1,800. We settled into the water. When we were two or three feet into the water, the rpm came back up to about 2,350 and we got airborne again to about ten feet, but with a very heavy one-to-one vibration. We were just able to stay up, and Commander Reed told the crewmen to jump out. Only one of the two crew had on a headset; he jumped out, and the other remained onboard.

We again lost rpm, and hit the water somewhat harder, with water splashing up on the windshield. This time we recovered full rpm, about 2,900, and had climbed rapidly to about twenty feet when the engine quit. This time we hit hard, with the rotors immediately contacting the water and rolling the helicopter to the right.

I got out of the helicopter as it sank. It settled on its right side, with the engine weighing the nose down, in about fifteen to eighteen feet of water. Commander Reed surfaced in about thirty seconds, but our crewman was missing. I found the cabin window escape hatch, four or five feet underwater, stomped it out, and was just ready to start feeling for the crewman when he surfaced. Some fifteen or twenty minutes later, we, including the crewman who had jumped, were picked up by a CG boat and later transferred to a civilian small boat that took us to Pinellas Point. A television reporter had started to take us back to the Air Station when the crewman who had been trapped in the helicopter passed out. We took him to the hospital emergency room, and continued to the base.

In summary, all of the B-47 crew survived. Two were in helicopter 1343, and one in the sling being hoisted, when 1343 crashed. The Air Force and Coast Guard crews were picked up by a small civilian boat. Our crew only had minor injuries, with the most painful being chemical burns from the JP4 and AvGas in the water.

Later, I was assigned, along with Lieutenant Martie Kaiser, to go to CGAS New Orleans, pick up one of their H-34s, and ferry it to St. Petersburg. We test flew on 29 and 30 September (had main rotor vibration and put on new blades), and flew CG 1342 to St. Petersburg on 1 October 1960. One of St. Petersburg's H-34s was subsequently lost offshore in a similar circumstance, and a crewman died.

I know of at least seven cases where the CG H-34s had an unexplained loss of power. In each case the air was heavy/moist, and the helicopter had low forward speed with normal power. In no case was the helicopter heavy, or approaching max gross weight.

Gerald Bickford, United States Army:

My assignments in Army Aviation were, overall, good, but the most enjoyable was my time working on and flying in the Sikorsky H-34. Like most crew chiefs, I was checked out on engine run-up. This was for engine maintenance checks, and to have the engine warmed up to save time for the pilots (they didn't have to wait as long to engage the main rotor). We also, on occasion, engaged the rotor and flew the helicopter. Believe it or not, a lot of crew chiefs were good on the controls.

One day I was on the flight board for a local flight with Mr. Jones and, as my bird had an hour-and-a-half to go for the

SP6 Gerald Bickford with Italian Sergeants Major Lonetto (left) and Longo (right) in October, 1962. The aircraft are CH-34Cs 57-1700 (c/n 58-844) and 57-1703 (c/n 58-847) of the 110th Aviation Company. (Photo courtesy of Gerald Bickford)

periodic inspection (one hundred-hour inspection), I knew it would be the "idiot circle" (take off, go around the airfield, land, and do it again and again and again). Sometimes we would fly practice GCAs (ground control approaches), which would get rid of some of the boredom.

After I had the H-34 ready, I got another crew chief to stand fireguard for me and started the engine. I then went through the run-up procedure. When I saw Mr. Jones coming, I reached up and released the rotor brake, and was going to shut the engine down but he was indicating to me to keep it running. He looked at the record book, closed the cabin door, and disappeared around the nose of the helicopter. The next thing I knew, he was climbing in the co-pilot's window, and I thought, "Well, I'm the brake man today." The toe pedals for the main wheel brakes were only on the pilot's side.

I removed my helmet from the co-pilot's side and put it on. After Mr. Jones had strapped in and looked over the instruments, he said, "Can't go anywhere if we're not engaged." I asked if he wanted me to engage the rotor, and he wanted to know if it was my first time. When I told him I had done it before, he said, "Let's do it." The engagement went smoothly, and after the preflight checks and tower contact Mr. Jones said, "Brakes off." I released the brakes, and as we moved forward out of our slot and cleared the other H-34s he said, "Tail wheel." I unlocked it. He made a left turn toward the maintenance hangar and the ramp which led to the taxiway, slowed down using the cyclic, and bottomed the collective. As the H-34 settled down on the struts, Mr. Jones told me to taxi out to the runway. This I had never done. How hard could it be? I heard, "You got it." I replied, "I got it," and then again heard, "You got it." Wrong. The H-34 had me.

As I pulled up a little on the collective and forward on the cyclic we moved forward. Just as I thought that this wasn't too bad, the tail had another idea and wanted to go right. I pushed right pedal and started to look at the grass. Too much. Push left, push right, crap—it's supposed to go straight. Too fast. A little brake, off the brake, a little pedal, uh-oh the turn through the ramp.

As we snaked our way past the hangar and the tail pylon waved good-bye to the aircraft behind us, I turned right to the taxiway. A little brake, right pedal, whee—too fast, fight to get it straight and then I bottomed pitch, neutral cyclic, brakes. I'd had it. Mr. Jones said he had the controls, and as I released the brakes and relinquished the controls happily we continued taxiing like he knew what he was doing. I think I heard a sigh of relief from the H-34 as we lifted to a hover, made a pedal turn to the left, and lifted away from the ground. Taxiing gets easier as you do it more often, but the first time was a fight I lost.

One evening, after flying Major General Fisher, Commanding General of the Southern European Task Force, we made the approach to Camp Pasalaqua, turning right from the river and dropping into the quadrangle. After we hovered to the center crosswalk, rattling the building windows and

drawing the usual crowds by the buildings, and set down, I unbuckled my seat belt, removed my headset (no helmets at that time), opened the door and got out. When General Fisher got out, he said, "Thank you," (he always thanked the crew at the end of a mission) and then asked me if he was alive. I said he was, and then he remarked how relieved he was and walked off. I looked at his aide, a Captain, as if to ask, "What was that all about?" He pointed up at the pilot, and I turned to look. The two stars hit me right between the eyes. The star plate was upside-down. I never made that mistake again.

The first time I flew with Mr. Schwartz, he asked for my maintenance manual. I didn't figure this out until after the flight. Because he was short, he would put the manual under the back of the seat cushion to raise him up. He also cranked out the tail rotor pedals and raised the seat. I don't remember who thought of it, but two blocks of wood were given to the crew chiefs to put under each seat for him. A few of us would check to find out what position he would fly, and set the seat up for him. Other times, I would ask him what seat he was taking and set it up, along with cranking the pedals, while he preflighted the helicopter. The pieces of wood became known as "Elmer's Blocks".

One Captain always wore highly shined shoes and boots. He would put wooden pegs into the foot step panels, holding them open so he wouldn't scuff the toes of his shoes while climbing up the side to the main transmission deck. Sometimes, while he was busy looking over the rotor head and transmission area, I would lay one of the pegs down and close the panel. Then I would be somewhere else while he tried to get his foot into the step with as little damage as possible to the polished footwear. Sorry, Captain. No disrespect was meant by me or the others. It was just something to do at the time for our amusement.

In all my years with the H-34, I only had about ten incidents with a droop stop remaining out during main rotor shut down. I didn't like it, especially with any wind at all, because the rotor blade would pass a foot or so over the tail cone. When the pilot chops throttle, two things happen at the same time. The engine goes to idle (1,100 rpm) and the hydro-mechanical clutch disengages the engine from the main transmission. Now that no power is driving anything, everything slows down and stops (except the engine). The bottom of each droop stop was painted red for easy viewing, and when the throttle was chopped I would hold my thumbs out, indicating to the pilot the droop stops were out. As the main rotor slowed down, and centrifugal force was no longer a factor, the droop stops would move inward, locking the main rotor blades and preventing them from moving below mid-travel (vertically). When all four were in, I turned my thumbs inwards and the pilot continued a normal shut down.

What to do if a droop stop didn't go in? Cursing didn't help, although I always felt better after saying a few choice phrases. As soon as I saw three went in and one stayed out, I signaled the pilot by keeping my right thumb out and moving my left forearm up and down, indicating to him that's what

I wanted him to do with the collective. At this time he was probably seeing a blade drooping below the path of the others. Sometimes this worked and the droop stop went in, and I would give the thumbs-in signal.

If this didn't work, I moved my right hand in a circling motion, telling the pilot to re-engage the rotor. After rotor engagement, and while we waited the two minutes for the oil to drain from the clutch, he played with the collective a little. Then we'd try the shutdown again. Time consuming? Yes, but sometimes it worked, and it beat the next procedure in my showing that obstinate droop stop who was the boss.

The last step in the "stubborn droop stop" episode was to let the pilot know it was out and continue shutting the aircraft down. The rotor brake was slowly engaged and, when the rotor stopped, released. Now my work began, and this took two people. The pilot would rotate the main rotor head, bringing the low blade over the tail cone, and I (standing on the stabilizer) would bend down, lift the blade to my waist, reposition my hands, and lift the blade over my head. Then I would give a little jump while the pilot had hand pressure on the droop stop. Bingo—in went the stop and the blade was where it was supposed to be. Man had, once again, beaten the droop stop.

Once, on final approach, I was told that both main rotor indicators went kaput. Two instruments don't quit at the same time, so I figured the rotor tach had sheared its shaft. Wiring didn't even come to mind in this situation. After discussing the problem, the decision was made to swap the rotor tach with the engine tach and, if that took care of the problem, continue home. A working engine tach is no good when the engine fails and you go into autorotation. That's when you want to know what the speed of the main rotor is. Too slow, and you fall out of the sky. Too fast, and you over-speed the complete power train. If you don't regain control, parts might take thoughts of their own, and—well, you get the idea.

I removed the rotor tach and, sure enough, the shaft was sheared. I removed the engine tach and installed it on the main transmission case. While I installed the cannon plug and did the safety wiring, one of the pilots installed the broken tach on the engine case and put on its cannon plug. I then safetied that tach as well. After the engine start, we were ready for the "play it by ear" mode. As throttle was increased, I listened to the engine, and when the sound was right (you know the sound after a lot of engagements) I signaled for rotor engagement. The pilot agreed with me, and turned on the clutch pump. As the rotor started to turn, throttle was increased, bringing the rotor speed up for the positive engagement. I knew the rotor tach was working, or he wouldn't have continued to engage. As the rotor kept picking up speed, throttle was chopped and then rolled back on. Got it.

The work was signed off, and I had one of the pilots sign as inspector for the grounding write-ups. I wrote up a one-time flight for engine tachometer shaft sheared. I don't put this incident in the class of "gethomeitis", but, rather, doing what had to be done in that situation at the time.

I don't know why, but I enjoyed sling work. I never used the floor window, because I could see a lot more lying on the floor and leaning out the door, and therefore could do a better job for the pilot. No matter what the pilot did or what type of situation might occur, you never changed your voice—always clear, calm, and slow. "Forward. . .forward. . .forward. . .right. . .right. . .hold. . .hold. . .hold. . .up. . .up. . .clear, you got it." You never went, "Forwardforwardforward. . .Okstop. . .rightrightstopleft." The pilot would go nuts. Even when it got hairy, I stayed calm. If you're putting a generator into, or taking one out of, a trailer, and the H-34 floats to the right and starts to drag, or turn the generator over, and it gets close to the side—don't worry about it. The pilot does not need to hear you screaming that you're going to crash and burn. Just keep the calm voice and move him to the left, bringing the aircraft back, level the generator, lift it out, and take off. I also changed commands prior to the pilot getting where I wanted him, because he couldn't change direction on a dime. If I wanted him at a spot left of where we were going over the load, as we were moving forward I would say "left" before my spot, and he would continue forward a little and then left where I wanted him. I never used the word "stop". The pilot knew that all my commands were to get him over the load and low enough to get hooked up. When I said "hold" he knew he was there. I liked low hookups because it was easier for the hookup man and therefore we had a faster load turnaround. I might have tapped a couple of guys in my low hookups, but I never squashed one.

A good sling pilot could help the crew chief by putting the right main wheel left and behind the load, and then move forward and right over the load. This would put the hook just about where you wanted it. When they did this, it made for some fast hookups, although it took away some of my fun.

Don Alberts, United States Navy:

I joined HS-4 in May, 1961, and shortly thereafter we deployed to Westpac and Southeast Asia aboard the USS YORKTOWN (CVS-10), being engaged in ASW and transport of Marines into and out of the jungle. I left during late Summer, 1962, for assignment as a special weapons instructor at Sandia Base in Albuquerque.

There were five West Coast HS squadrons: HS-2, HS-4, HS-6, HS-8, and HS-10 (the RAG squadron). There was very little interplay among them, since we tended to be deployed in rotation, and seldom got to see the other people. HS-4 tended to see itself, and I think be seen by the others, as the premier squadron, probably because it was the Navy's first all-weather, night-dipping helo squadron (known as the "Black Knights"). Certainly, the Skipper always thought so, and we did more genuine black-night flying than I could have imagined possible. This resulted in the "E" that is seen on our planes' sides in many photos, and also really cut into my movie watching. For ASW work, we carried both active and passive homing torpedoes as well as conventional depth charges and the atomic depth bomb (or at least the dummies of it).

HSS-1N #61 (148003, c/n 58-1255) of HS-4 preparing to launch from Yorktown's flight deck, October, 1961. (Photo courtesy of Dr. Don Alberts)

A Seabat pilot's view of USS Bennington (CVS-20) and HSS-1Ns of HS-6 in Tokyo Bay, February, 1962. (Photo courtesy of Dr. Don Alberts)

As with most pilots, we claimed to like the aircraft we had to fly, but as an aero engineer I considered it quite underpowered and complicated, therefore vulnerable to outside influences. Its magnesium skin was a source of concern to some extent, since it would burn like a flare if ignited properly, as one H-34 was when the early landing gear caused deck touchdown problems on the YORKTOWN. The vibration simply tore the plane apart, shedding burning parts in all directions, with the pilots miraculously surviving at the center of the chaos. The landing gear problem was fixed by the time I joined HS-4, but everybody remembered or knew of it.

All HS-4 planes were in the engine gray and red-orange (not dayglo, as far as I can tell) color scheme, with "NU" as the tail designator and a medium blue tail rotor transmission housing. All rotor blades were painted very light gray on upper surfaces and matte black on undersurfaces. That was so that, during preflight inspections at night, you could quickly check to see if the rotor blades were properly set. If you could see a white blade, it was on upside-down and would ruin your whole night. I had that happen once while flying with the Exec, a careless hard-charger who wanted to get into the air fast. The system really did work, and saved planes several times.

The Doppler radar and radar altimeter were integrated into the automatic night approach and dipping control, which as far as I know HS-4 pioneered. It was a source of constant trouble and anxiety, and we trusted it very little while lowering us in an inky night to a very low hover. We tended to override it manually at the least little indication of malfunction. The radar altimeter was useful over the ground terrain, however.

The engines were pushed to their limits as powerplants. Over-boosts and over-speeding were common. I was the maintenance officer for powerplants, transmissions, clutches, etc. (as the only aero and mechanical engineer in the squadron, even though a junior officer). It was a very demanding

LT(jg) Jim Biestek (left) and LT(jg) Don Alberts with HS-4's #61 aboard Yorktown in November of 1961. (Photo courtesy of Dr. Don Alberts)

job, but one in which I took great interest, since flying a single-engine plane over either ocean or jungles will focus your attention on such things. I still have a great affection for the H-34, but that doesn't mean it was without fault.

Al Kerst, United States Marine Corps:

I was a 6611 (aircraft radio/avionics technician), and also helped tinker with everything on the UH-34D except the engine, hydraulics, and the transmission, power, and rotor areas (except to lend manpower to the mechanics, metal-benders, crew chiefs, etc.). I was familiar with the old tube-style ASE, which gave the "camel drivers" fits from time to time.

I also helped with the very first M-60 door mounts done in the field. Our metal-benders actually took some three-quarter inch steel scrap bars, bent them with heat, cut up some old machine gun tripods, did some welding, sanding, painting, and door mounting, and presto—we had a door mount that would swivel left, right, up, down, but yet was limited in swing in all directions to prevent accidentally blowing off landing gear, tail rotors, or main rotors. It was effective, and after a little adjustment we installed it on all the "Dogs" we had. This was at Da Nang, Operation SHUFLY, in September of 1962. Prior to that we "strap-hung" in the door or out the left rear window with any automatic weapon we could scrounge up: an AR-15 Armalite, M-3 .45-caliber "Grease Gun", and even used Thompson .45-caliber submachine guns with both disc and tube magazines just like Al Capone or Machine Gun Kelly. Eventually we ended up with AR-15s in the rear window until the M-16s arrived. We even tried the M-14 on auto but it was too darn long and unwieldy. Later on (after my time there), crews even used grenade launchers and rocket launchers out the door and rear window—anything for effective firepower.

We were the first unit to make hot LZs not so hot with immediate suppressing firepower. It caught the VC by surprise at first, and then they changed tactics. We even had one co-pilot who was trying to "synchronize" his .38 through the rotor blades when a sudden sweeping, banking turn to the left was made. Yup—he hit 'em! The Maintenance Chief and various other dignitaries lit into him like the blazes when they got back. In fact, the crew chief (an E-5), met him as he was climbing down the side, got in his face and said, "You dumb shit!" and then walked away.

I believe our unit also suffered the first wounds and KIAs for Marine air crews. In late 1962, we had a crew chief who had a round come up through the floor, through the web seat and flak jacket he was sitting on, tear a hole in his flight suit and skivvies just off to the left of his normal "Brown Spot" without breaking his skin, and then bounce to the floor and spin between his legs and out the door. You have never seen such a welt on a guy's ass! I do not believe he got the Purple Heart for it. The Flight Surgeon who treated him for the "almost wound" was going to certify it for the PH but he himself became one of our first KIAs just a few days later. The KIAs came when weather closed in suddenly on a hot LZ. The third "Dog" in line flared out into the closing weather, and as it turned out there was a mountain inside the weather. They are all listed on "The Wall" tablet that starts with 1959, and were the first USMC air crew killed in action in Nam. The records say "non-hostile". That is disputed by many of us who were there.

On the first wave in, the LZ was cool. The VC were lying low. On the second wave, it got hot. The third wave took a wave off because of the VC action. The weather had closed in and the third bird (last one) in the third wave flared out into the weather and then into the mountain. The pilot survived.

The co-pilot was Lieutenant Tunney. The crew chief was Corporal Anderson. The window gunner was Sergeant Hamilton.

This last bird in was our Crew Support bird with various maintenance supplies and of course "the Docs". Lance Corporal Valentin was a 6611 Avionics Support Tech and Sergeant Pendel was the AvGas Refueling Specialist. HM2 Norton was Squadron Corpsman and Lieutenant(MC) Griffin was the Squadron Flight Surgeon.

I was on the recovery crew. Anderson could not be recovered due to the severe fire. The rest died from burns and various injuries during the night on the side of the hill. Dr. Griffin was a hero. He attempted to treat the rest during the night but lost so much blood that he died in the process. Myself and Gunnery Sergeant Smith, Squadron Administration Chief, had to go out and do the Graves Registration bit (fingerprints, dogtags, eyeball certificates, etc.). Hamilton and I were bunk mates. I was on this bird, but Valentin came out from the line shack and yelled at me that Master Sergeant Harvey (Avionics NCOIC) needed me for something else that had come up. I swapped out with Valentin, and he is there and I am here. Semper Fi!

As far as the bird itself, she was noisy, she shook and rattled, but she was good! By the way, our mechs were the first to change an engine in the field (a rice paddy on the side of a hill in hostile territory). We flew the old engine off, flew in a new one, hooked it all up, ran it up, and flew the "Dog" home.

Luther Stephens, United States Marine Corps:

I really enjoyed my H-34 time. That's what I cut my teeth on. At the time, the Marine Corps was building up the helicopter community. I was a heavy equipment operator at Camp Lejeune, and I retrained as a helicopter mechanic at Memphis, Tennessee, in the Basic School there. The aircraft they were using in the school at the time was the HRS-3. We didn't fly any at the school, but all of the training maintenance evolutions were done on the HRS-3. It had a lot of similarities to the H-34, but the H-34 was a vastly improved machine.

While I was at school, before we completed our training, they had received a HUS-1 for the school to start training mechanics. Part of our task, while we were waiting for assignment out of Memphis to squadrons and had a week or two to play with, was to help disassemble the helicopter that was received to be used as a trainer. We cut the rotor blades off so they could actually turn the rotor while it was in a shop somewhat smaller than what the true rotor diameter was, but left the blades intact at the hub so they could demonstrate blade removal.

That was early 1962. After school, I was assigned to MCAS New River. MAG-26 was the only MAG there, with four squadrons attached. There was a lot of competition between squadrons as to who was best. Everybody was in stiff competition for the most flight hours every month. I can recall, with twenty-four aircraft in a squadron, if the squadron wasn't getting two thousand hours a month it wasn't hacking the load. That seldom happens these days.

Not too long after I got to MAG-26, there was a Group change of command ceremony. They imposed quiet hours—there were to be no local flights, so the change of command ceremony could proceed without any interference from aircraft. I was in the duty squadron, and had to maintain a ready crew in case somebody had a medevac or some other kind of problem, so we weren't a part of this large formation out on the flightline. We had to take a blade off of one of the H-34s. With the quiet hours on, wouldn't you know that's when you're going to goof? Somebody slipped, and we ended up dropping the blade. When the tip hit the ground, those steel spars made such a tremendously loud clunk that it was heard throughout the formation that was being duly impressed by the ceremony. That lead to some really big ass-chewings for us maintainers. I guess that's one way to get attention if you think you're being left out of something!

When James Meredith enrolled in the University of Mississippi in the Fall of 1962, when the big blowups about segregation and various problems like that were occurring in the South, because of the anticipated problems down there in Mississippi, Marine Aircraft Group Twenty-six was put on alert. I got a knock on my door one Saturday evening about eight o'clock. There was an MP there, who said, "Pack enough stuff for four or five days, and report to your work department at the Air Station." So I packed a bag and headed over to the flightline, and when I got there I was told, "Go down to the armory, draw your rifle and your flight gear. Preflight your plane and sign it off. We're going someplace." I did, and we did.

We left the Air Station that night; HMM-262 was one of the first squadrons out. When we left it was dark, and when we landed after about a four-hour flight I didn't really know where we were because it was still dark. We parked and shut down, and finished the night in the helicopter.

We woke up in the morning, got out and looked behind us, and it looked like every helicopter in the world was there. We were parked on a couple of old runways down near Atlanta, maybe at Dobbins Air Force Base. We refueled and got some chow, then launched out of there and headed over to Memphis.

We set up camp—and I say "camp" because that's literally what it was—on the grassy infield between the runways.

The entire MAG was there. All four squadrons were represented, and just about everything that was flying at New River showed up. The 82nd Airborne was flying into Memphis in C-130s, landing and taking off at the rate of three a minute or so. It was a tremendous build-up.

We ended up being down there for about two weeks, and we operated out of our back pockets. But that was one of the great things about the H-34s: you didn't have to have a truckload of support equipment following you around. We could maintain it pretty much from our toolbox, with just some gaskets and bits and pieces, and a new radio every once in awhile.

We were ferrying troops around at whoever was running the show's beck and call, moving them wherever they thought the hot spots were going to be, to be used for riot control. The day of the big event, there was a homecoming football game. They loaded us up with sticks of six or eight Army guys, all carrying riot gear and stuff like that. We flew *en masse* over to another Air Force base, within a half-hour drive from the suspected hot spot. We occupied one of the runways, because they flew the whole gaggle—there must have been a hundred helicopters—in there. We shut down, and we sat there and waited and waited and waited. . .

Just about dark, they suspected that nothing was going to happen and everything was under control. We had the luxury of starting engines, engaging rotors, flying in a big daisy chain around the base, and re-parking. We were blocking one of the runways, and the Air Force wanted us to park in a different area. It was kind of neat, because the sky was just about filled with H-34s. For a young fellow (at the time) like myself, it was exciting.

The Army guys were carrying, amongst their riot gear, flame-throwers loaded with tear gas. Occasionally, these tanks would develop a leak. That would get kind of exciting, when flying around, to have tear gas escaping and swirling around in the cockpit. As a result, we got to fly a couple of times with our gas masks on.

We never did "go to war" over that. They calmed things down without the intervention of some eight thousand troops. We accomplished our mission and returned to New River.

Our squadron also participated in the Cuban missile blockade. We were still stationed out of New River at the time, and we had several aircraft embarked aboard USS THETIS BAY (LPH-6). We were assigned to her for carrier qualifications. It was supposed to be a leisurely two-week cruise, during which we'd get our pilots carrier-trained. As it turned out, the balloon went up down south and we ended up embarking the rest of the squadron for that deployment. Fortunately, we spent more time in port than at sea due to the immense load that the ship had on it; that ship wasn't really equipped to provide for the embarked people: a very old ship, very small, and highly overtaxed for the mission that was assigned to it. We survived that part of it, and finally did our offload to Camp Lejeune. About mid-day, while we were ferrying troops to Camp Lejeune, we came back to the ship after dropping off our load of troops. While we had gone out and back, the ship had changed course. They were making steering speed along the coast, sailing very slow in a racetrack pattern. What little wind there was caused us to be downwind. The pilots weren't expecting it, due to a lack of communication or something. The conversation in the cockpit went something like, "Did you see that over-boost?" The response to that was, "No I didn't. Crew chief, did you see that over-boost?" "No, Sir, I didn't see an over-boost." We picked up our next load of troops, and all the gauges looked fine, so the pilots decided, since it was a short hop to Lejeune, we'd try it. We didn't want to spend another minute aboard ship; we were pretty sick and tired of it at that point. We proceeded back to Camp Lejeune, dropped the troops off at the main parade ground uneventfully, and during the subsequent liftoff the chip light came on. About the same time, the engine coughed once and quit. It can get awful quiet in one of those aircraft when the engine quits. We were a couple hundred feet off the ground, and we made a quick descent into a vacant parking lot. The pilots landed it very nicely, and we shut the rotor down. I got my trusty toolbox out. We opened up the engine bay and pulled the filter and chip detector as was SOP. Both of them had sufficient material on them that we knew we weren't going anywhere with that engine. The pilots called the Air Station, and they said they'd send out a crew with some blade saddles, and a truck to tow it home. The truck arrived late that afternoon. We folded the blades and we hooked up the tow bar. We rode behind this 6x6 truck, towing the H-34 some fifteen miles from mainside Lejeune out to the airfield at speeds below five miles per hour. It took a significant amount of time, clearing branches and wires over the road. We were among the last in the squadron to arrive. Almost everybody else had shut down and secured for the night by the time we got there.

I think it was in the Spring of 1964 that HMM-261 had to provide VIP transportation for Robert Kennedy. At the time, the United States made a substantial contribution for the restoration of Corregidor. There was a pretty big publicity splash. We sent a four-plane flight to Corregidor, and we spent the day looking around at all of the historical material there, in the camps and so on. We flew several support missions for the restoration from Manila.

After I came back from one Westpac cruise, I got orders to HMX-1, and worked in the H-34 section at Quantico. It got pretty small pretty fast, because the H-34s were starting to be replaced with H-46s. We had a lot of veteran H-34 pilots in HMX-1, and I got to serve with some of the best in Vietnam later.

In Da Nang, we were flying a required test flight on my aircraft after a major inspection. The crew chief always had to fly as the co-pilot on a test flight, so we got to do some cockpit flying whether we liked it or not. We flew around for awhile, and Herb East, the test pilot, put it through its paces. In straight and level flight, he shut off the auxiliary servo, which controls the tail rotor. With the servo out, you're still supposed to be able to control the tail rotor. He tried to move the rudder pedals, and couldn't. He was practically standing on the rud-

der pedal with one foot! He said, "I don't believe this! You try it. Give the rudder pedal a push as hard as you can." So I did, and I couldn't get it to move. The tail rotor would not change at all. If we turned the aux servo on it would work. We put her back down on the ground, disconnected the pitch links, and found all of the pitch-change bearings in the hub had become so clogged with dirt from our flight ops that only the power of the aux servo, with its hydraulic boost, would overcome the stiffness in the bearings. With flight loads, without the boost on, there was nothing you could do to control the tail rotor. We replaced the hub, needless to say, and it was fine after that. That was one of the things that we started checking for quite frequently after that, as a way to survive.

I can't relate a lot of war stories, but if I remember I was one of the high-time crew chiefs. I had something like five hundred hours in the four months that we were there. Our first section seemed to do real well in that regard. We always had our aircraft up, and were always available and on the schedule. So, we got a lot of flight time. That was good. We

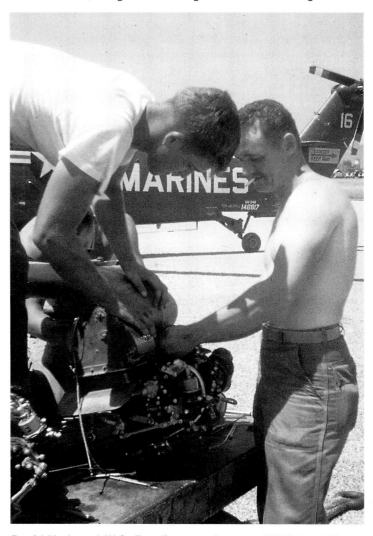

David York and W.S. Bendixen work on an HMM-163 UH-34D carburetor at Da Nang in 1965. The Seahorse is 148817 (c/n 58-1406). (Photo courtesy of David M. York)

got shot at occasionally, but we had more fear of the Vietnamese that we were carrying around leaving unattended hand grenades and things like that lying around in the cabin after a troop lift. After one emergency lift we made—we lifted a bunch of ARVNs into some landing zone somewhere—we came back really late at night, so we just shut the aircraft down, tied them down, and hit the sack. The next morning, we started looking over things in the daylight, and we found several loose grenades in the cabins of the helicopters.

Whenever the ARVNs went to a village, they carried their supplies live. Live pigs, live cows. They had no refrigeration, so the only way to have meat was to have it on the hoof. We got tasked many times to carry live animals. Usually the pigs were in thatched cartons, made out of bamboo grass and stuff like that. Chickens were handled the same way. But cows wouldn't be in any kind of container—pretty tough to do with a cow—so they'd just be hobbled so they couldn't walk around. We'd tie them into the cabin. We were in a two-plane formation, flying wing on a buddy of mine who had a couple of cows in the cabin. Evidently, one of the cows got loose. We heard some conversation on the radio, and the next thing we saw was a cow spiraling out of the door at two thousand feet. Sometimes the pigs would break loose. They were pretty excited critters, and would start running around the cabin. They were always able to find the door.

Max Mitchell, United States Army:
In 1963, the town of Longarone in northeastern Italy was completely devastated when a portion of a mountain slid into the lake behind the Vajont dam, causing a wall of water three hundred feet high to go over the dam. The dam was a thousand feet above the town of Longarone. The engineers said that the air displaced by the water is what caused most of the damage.

I was in the 110th Aviation Company in Verona, Italy. I took three H-34s to Longarone about seven-thirty the morning of the disaster. Ground support followed, and more aircraft and H-34s were brought there. We stayed until the Italian Aviation arrived about ten days later. They had nothing like the H-34, only small helicopters. The 110th was in the world news; the H-34s were angels of mercy. Reporters came from all over the world.

During the relief effort, I was sitting in the cockpit of my H-34 at Longarone when I saw some men in a small boat trying to get across the fast moving river. It looked as if one man in the boat had a rope. When the H-34 was unloaded I flew over to where the men were and set down. I got out and walked over to the men. In my limited Italian, I asked what they were trying to do. They asked me to wait, and shouted to another man about fifty yards away. He came over and said, "Hello," in English. I asked him the same question. He said they were trying to get the small rope over to the other side so they could pull a bigger rope over, then the boat with some men. Next they would pull the big steel cable needed

to put in a floating bridge so they would be able to move medical supplies and food to a small down river about fifty kilometers. The cable was about an inch in diameter. He said they had been trying for hours to get the rope across but the water was so swift it would carry them downstream too far and the rope would play completely out.

I called my crew chief over and discussed with him the possibility of pulling the cable across with the helicopter. He agreed it was indeed possible, so we went back to the chopper, checked the hook, and talked a bit about the pros and cons, what he would do and what I would do. We went back to the Italian engineer and told him we would take the cable across. At first he didn't believe it, then his eyes lit up and he told the others, who broke into big smiles, cheered, and shook ours hands.

We took five or six of the Italian men in the helicopter with us. The crew chief had rigged the hook and shown one of the men how to attach the cable to the hook. I flew the helicopter above the cable, and the Italian hooked us up. The cable was on a big wooden roller that rolled easily on an axle. I eased in the power and started moving forward with no problem; it did not seem to be any load at all. About half way across the river it began to pull a bit, so I applied more power and eased the nose down more. When I reached the other side of the river I was just about at max power and extremely nose down. I finally added max power and got the cable about a hundred feet up on the shore. Now I had to lower the aircraft and back up a bit to get slack in the cable before I could release it. Everything worked out just fine, and with white knuckles I set the chopper down.

Needless to say, we had some cheering, laughing, dancing fans. That was all the thanks we needed.

I asked the English-speaking engineer if the town needed supplies now, and he said they did. We loaded all the medical supplies and food on the helicopter, and ten minutes later we were unloading them at the village.

That evening our coordinating team, headed by Captain Jim Stamper, arrived. I talked to Jim about the town's plight, and he agreed to furnish daily support to the town until the bridge became passable. Every morning we made the short flight to supply the town. We were always greeted, mostly by women with smiling faces, occasionally tears and a smile. It was extremely heartwarming to be able to help. The people were always so appreciative.

At Longarone we carried anything or anyone that needed to be evacuated: men, women, children, goats, pigs, dogs, calves, cats, and canaries. The town of Casso was above a dam on the edge of a lake. The Italian authorities feared the town might slide into the lake. They told the people they had to evacuate. Three helicopters were sent up to evacuate them, which we finished about five in the afternoon. The next morning they were all back. The authorities told them again they had to evacuate, and again we evacuated them. The next morning—you guessed it—we evacuated them again. They were coming back to get things they needed, and we moved their things for them. It slowed down after they had most of their belongings with them.

All of the American Army pilots were awarded large certificates, with the disaster scene watermarked on it, and medals by the President of Italy.

Years after I retired I went back to Longarone. It had rebuilt and was very beautiful. I picked up some pictures of the disaster. It was then that I discovered that 3,700 people died that morning.

Here are a few other things the 110th did in Italy with the H-34s:

Put a statue of Gabriel blowing his horn on a church steeple in Verona.

Built *refugios* (refuges) on numerous mountain sites all over Italy.

Provided medical evacuation of Alpine troops.

Looked for lost or missing mountain climbers.

A heavily damaged UH-34D from HMM-163 is recovered after ditching because of engine failure. (Photo courtesy of David M. York)

Dodged rain rockets, used to seed clouds.

"Rain rockets" were small rockets with some type of chemical in them to (for lack of better description) milk the clouds of moisture. I was in Verona, the Chianti wine region. The wine growers would fire these rockets into clouds, and the chemicals would develop or cause the clouds to give off rain. I'm not too sure how effective it was. We were supposed to get notification when and where they were going to fire the rockets, so we could stay clear of those areas. Often, especially in the afternoon, some grower would fire off the rockets in an area that was not on the daily list we received. If we saw them exploding to the north, we would go south. Even when they were firing where they said they would, it was unnerving when you were flying to see explosions about a half mile from you.

The only engine failure to occur while I was flying was due to an over-speed. 2,800 rpm was the max the Wright R-1820 could handle. Pilots were required to report to Maintenance anything over. 2,801 was an over-speed and the sump strainers had to be pulled. Over 2,900 was called a "Downing Discrepancy", and the engine was replaced. Because there was no prop, the engine was easy to over-rev on startup. Also, on takeoff and hover, I believe max power was 2,800 rpm and fifty-six inches of mercury, so one would have to be very careful with the throttle. Sometimes I think there were too many distractions for a pilot to notice a slight over-speed, but almost every sump I pulled resulted in metal flakes.

Ed Kozak, United States Marine Corps:

I first became aware of the UH-34D when I arrived at my first non-training duty station at Quantico, Virginia. The squadron to which I was assigned was HMX-1, commonly known as "Marine One". At Quantico, we called the H-34 the "Work Horse". In this peacetime environment, we spent as much time polishing our assigned aircraft as we did on pre- and post-flight maintenance. The usual rule of thumb then was about one man-hour of maintenance for each flight hour. In other words, if the plane flew for two hours, one might expect two men to perform around an hour's work afterwards, including refueling, greasing, and clean-up. The main rotor head, I believe, carried sixty-eight grease fittings, the tail rotor another twelve. Lots of greasing, usually after each five flight hours. Our squadron used these aircraft for sea and land rescue, as well as transportation of Executive Office personnel from the President and Chiefs of State to the Secret Service (and press people from time to time). We were, in fact, the support squadron for the White House, and HMX-1 remains in that role to this day. When I was stationed there, the Marine Corps worked hand-in-hand with the US Army, who furnished the President with the same military service as did the Marine One boys. There was a lot of competition between the two service branches, but the Marines, as usual, took the lead. Today, the Marine Corps is the sole provider of helicopter transportation to the President. HMX-1 also was responsible for testing proposed or newly-acquired Marine

Corps helicopters. While I was stationed there, we were testing the readiness of the brand-new CH-46A.

In mid-1965 I volunteered for combat support and direct combat flight operations in Vietnam, and I arrived in-country in June of that year. There, the H-34 took on a new, rather basic, "Plain Jane" appearance. In the warm climate, there was no need for a cabin heater. Removed, too, were all the crew (or cargo) compartment Plexiglas windows and the cargo door on most of the squadron's planes. They were not all spit-shined, and were a duller shade of Corps green. Helicopters in my first squadron, HMM-363, initially bore white markings and the squadron designation "YZ". Later, the lettering was changed to flat black. Likewise, aircraft in our sister squadron, HMM-364, bore the same color schemes but were designated "YK".

The UH-34D could take a lot of punishment. One of the biggest differences between her and the turbine powered UH-1E helicopters of Vietnam fame was that the UH-34D's radial engine could filter out most of the fine sand particles, while the Huey's turbine engines ingested the sand and suffered drastic reductions in engine life spans. The piston powered UH-34D used an air filtration system very much like an automobile's, and it could easily be cleaned and reinstalled whenever necessary. I don't recall any engine failures due to sand or foreign body ingestion during my tours of duty, which ended in November, 1966. However, the Seahorse did suffer from lower power and hauling capabilities in comparison to the UH-1E. Birds like the CH-46, and the even larger Army CH-47, could haul much, much more cargo and personnel; they were much larger than our favorite helicopter.

One of the biggest problems I encountered was having the engine over-boosted, along with the fouling of spark plugs from long engine idle rpm running time in secure or unsecured landing zones (LZ). This would often result in a less than acceptable rpm drop between right and left magneto checks. I believe the upper limit was a drop of twenty-five rpm from either both to right or both to left mag settings from the cockpit. After long periods of idle-engine rpm waiting for sorties, the plugs became carbon-fouled. If a power burn off didn't solve the problem, it became necessary to replace the plugs. Cleaning was not SOP, but on more than one occasion it was done as best as it could be done in the field just to get the hell out of harm's way. In the field, rarely could the crew chief remove, wire brush, gap, and replace all the plugs, as each of the nine cylinders carried two spark plugs and the back upper ones were almost impossible to get to in the boonies. On one occasion, I was part of a team that had to remove the engine in the field. We did it basically with a large screw driver and hammer. Most of the engine connections were the quick-connect/disconnect type and required no tools to disconnect fluid or electrical lines from the engine or other components such as the clutch or transmission. We loosened the damaged power plant and let it drop to the ground. Much

more care was taken when the replacement engine arrived along with proper tools, rigs, fixtures, and manpower to install it properly.

Initially, the armament we carried other than personal sidearms was a .50-caliber machine gun. These heavy guns caused fuselage cracks as they vibrated excessively when used. Many of the cracks were along station 235 (if I remember correctly) and were just stop-drilled to prevent the crack from growing any larger. Sometimes the metalsmith would pop-rivet in a few places along the crack and that was the main repair for such problems. In a short time the heavy guns were replaced with the M-60 machine gun which used 7.62mm NATO rounds, and this worked very well. Two were mounted, one at the crew chief's position immediately forward of the starboard cabin opening and the other forward of the first cabin window opening on the port side. We also carried various colored smoke grenades to be used to mark landing sites, note wind direction for the pilots, and mark sites of hostile incoming fire, and a few thermal grenades to destroy the plane if need be. We usually carried two hundred rounds of ammo per machine gun. Some crew chiefs prepared all tracer-round belts that were used to set fire to hootches and otherwise mark out an enemy site. At first, there was no armor to protect the plane or crew. Added was a pressed aluminum alloy oil cooler shield under the fuselage immediately to the rear (actually front as the engine was mounted in backwards) of the engine. Each of the enlisted men sat on a piece of boiler plate and his flak vest. The vests, at least in my case, were sat on rather than worn. It was a matter of personal choice, however. I just didn't relish the idea of my family jewels being taken away from me.

Some of the pilots, just a rare few, were not really good ones. Perhaps lacking flight experience is a better description. In the Corps, a crew chief, who was the senior enlisted person assigned to the helicopter, stayed with that particular aircraft and flew wherever she did, as did the second mechanic (who doubled as the port window gunner). The pilots, officer material, were assigned to different machines each day. A skilled crew chief often knew who did and who did not treat his chopper with great respect. An older, much more experienced crew chief than I told me that when I began crewing on my own (which I did shortly after my arrival in-country) to periodically look up from my seat in the cargo compartment, through the legs of the pilot. In doing so, I could see the manifold pressure gauge. The gauge measured intake manifold pressure in inches of mercury (Hg) and indicated whether or not the engine was being over-boosted, or given too much power considering the situation at hand. The reasonable or maximum inches of mercury was often dependent upon load, weather conditions and other pilot-only-needed-to-know stuff. All I understood was that if the gauge read much more than sixty-five inches Hg, we could expect some trouble, like sucking in a carburetor gasket. That definitely made the helo fade from altitude very fast!

Such was one experience I had the fate to suffer. We crashed at sea in early February, 1966, after the engine was over-boosted. We were carrying too much weight in coiled barbed wire, and the old bird just couldn't haul it. A hover check had been requested by Ron, the crew chief. It was not performed by the aircraft commander, who felt the chopper was capable of carrying the heavy load. He was mistaken. We pancaked into the South China Sea about two miles out early in the morning when it was still dark outside. The plane sank like a rock in about sixty feet of water. One man died (the crew chief—I was working as the second mechanic that day).

Once well-learned in the effects of enemy action, I was able to pinpoint where a round probably hit us just by the sound it made. For example, a hit to the main blades often resulted in a "one-per" vibration felt in the control sticks as well as throughout the airframe. By the type of vibration and the impact sounds, one could guess quite correctly much of the time whether it was a hit to the blades, rotor head, drive shaft, etc. The UH-34D was a sturdy helicopter, and the ones I flew in which ran into hostile fire just kept right on flying most of the time. In one major operation, one of our birds took over three hundred rounds without any injury to crew or passengers, and she kept right on flying all the way back home with little damage to critical components or avionics. Once, overnight at Qui Nhon, we were doing an engine inspection. Both the clamshell doors were secured open when one of the only two Army Sikorsky Flying Cranes then in-country taxied by us. We lost one clamshell door to the intense rotorwash and the other was badly damaged. When one of the helicopters was so badly damaged either through mechanical failure or hostile action as to be unflyable, it was not given away. They quickly became skeletons, as useable parts were cannibalized to keep the other planes going.

The sand of the coastal areas took its toll on blade tips, which were replaceable. It also made greasing the rotor heads a major task, as care had to be taken to wipe excess grease to prevent sand from accumulating on the fully articulating rotor head. Blades were seldom folded unless anticipated weather conditions warranted it. Many crew chiefs used a combination of hydraulic oil and AvGas to clean and give the fuselage a bit of a shine. Not to worry, just a touch of gas to lots of hyd oil. It was wiped dry so as to keep grime from being attracted to the airframe. Sand also provided the major reason why a small broom or hand broom was kept on board. Besides sand, blood and guts were another mess dealt with almost on a daily basis. Collecting spent brass was another chore I personally hated to do.

Crew chiefs sometimes slept overnight in the Sikorsky Hotel. Not very comfortable, but better than the rice paddy. In the field, the air crews often had the wherewithal to have hot food, albeit the infamous C-rats. We would take a used, cleaned-out, larger C-rat can, punch a few holes in the bottom and lower sides, fill it halfway with sand, and then add a few drops of AvGas (aviation gasoline: 115/145 PN, purple colored for identification) taken from one of the belly draincocks. At a safe distance away from the plane, we'd light the "stove". Grunts were always begging for one of the stoves we used on field trips. If we were in a sour mood, we might not let them use our supply of AvGas for their stove fuel, using the excuse that we didn't have much to spare (which was totally untrue).

The biggest advantage of being a UH-34D crew member was that you got to leave the LZs. The biggest disadvantage was that you flew into an LZ as a target over fifty-eight feet long, going very slowly at a predictable speed and direction. It was at times a real nightmare for both crew and passengers.

The most satisfying missions were the frequent medevacs that saved Marine lives, as we were seldom farther than fifteen minutes from an aid station or field hospital, or even a hospital ship like the USS REPOSE (AH-16). The most difficult missions were recovering the dead to take to the carrier for storage before their shipment back home.

The UH-34D was noisy, shuddered a bit and had its shortcomings. But even today, I would rather crew one of her kind than the newfangled techno flying black boxes!

Roger W. Ek, United States Navy:
I arrived, on 2 May 1965, on the first US ship to reach the Dominican Republic. It was the USS NEWPORT NEWS (CA-148). I had a single-seat UH-13P which had a bench seat for three behind it. I made the first flight into the Embassy compound. That afternoon, the Marines pulled down the bronze lampposts inside the compound so larger helicopters could land.

Lieutenant Con Jaburg flew the station UH-34D from Roosevelt Roads, Puerto Rico, to the NEWPORT NEWS off Santo Domingo on 3 May, because we needed a bigger helicopter than the UH-13. The H-34 was painted dayglo orange, as were most rescue aircraft at the time. I flew with Con when I wasn't flying the Bell. Our Embassy was attacked on the third. The only gunfire I saw directed at me when I was airborne was pistol fire from a bunch of men that jumped from a car.

I flew the UH-34D and the VH-34D out of Quonset Point Naval Air Station from 1966 to 1969. The VIP bird was used to fly guests to the Naval War College, and to fly Kennedys to their mansion at Newport. I also flew Vice President Humphrey to look at flood damage in 1968.

Leonard Martinez, United States Marine Corps:
In Vietnam, flight crews were routinely scheduled for early morning missions. When you drew these flights you were always awakened at three in the morning. This gave you time to wash up, grab some chow, preflight your bird, and be airborne on your way to your first LZ by sunrise. Every crew drew these missions on a rotating basis.

All enlisted squadron personnel were assigned specific hootches and cots. There were normally eight cots to a hootch. If you were assigned an early flight, you would be notified by your section leader at evening muster. It was each crewmember's responsibility to notify the duty clerk of his flight status and confirm where his sleeping quarters were located. The next morning the duty clerk would awaken you (usually by giving you a good shake and shining a flashlight into your eyes), then verify your name and early morning flight status. Once he was sure that you were wide awake, he would require that you verbally confirm that you had been awakened, notified of your early flight, and have you sign the wake-up sheet. At this point it was your responsibility to make your assigned flight. The first time you missed an early flight, you visited the C.O. for a good ass chewing. Your second offense was punishable by a court martial. I don't recall anyone ever committing two offenses, but there were plenty of first-time offenders. I never did miss an early hop, but on several occasions I did sleep-in an extra half hour, missed chow and flew on an empty stomach. Early morning chow was never worth losing some extra sleep over.

In mid-February of '67, our squadron was scheduled for a routine three month rotation to Okinawa. Since I only had five months in-country, the Corps felt that I had not earned the right to enjoy three months of R&R. I was therefore taken off flight status and transferred to HM&S-16, the air group's heavy maintenance and support squadron. My departure created a vacant slot at the aircraft crew chief position. My regular gunner, Lance Corporal Gary Nixon, was promoted to this assignment. This left an empty seat at the gunner's side. Since the squadron only had two active flight days before they departed out of country, it was decided to fill the gunner's position with a volunteer from the squadron personnel who typically were not assigned to active flight status. It was common practice to supplement the active flight crews with administrative and ground support personnel. There was never any shortage of volunteers. They all wanted to fly combat mis-

"The clear area on the ridge line was our landing pad. At the time I took this photo we were taking heavy fire from the VC, so we had to shoot an autorotation in." (Photo courtesy of Leonard G. Martinez)

sions in order to earn Air Medals and a combat air crewman's wings.

The day after my transfer, Nixon was scheduled for the afternoon medevac standby. Lance Corporal Jim Childers, one of the squadron motor transport drivers, was scheduled to fly as his gunner. Childers had previously been assigned to a grunt outfit, so he knew how to handle an M-60 machine gun.

Late that afternoon they got a call to pick up a WIA. En route to the LZ they were informed that they could expect some enemy fire. They made a routine landing and pickup, but as they were lifting off they started receiving a heavy volume of enemy small arms fire. The bird took several hits. Unfortunately for Childers, one of the rounds had his name on it.

Part of our standard flight gear included the wearing of either a flak jacket, standard issue for every grunt in Nam, or a heavier bulletproof vest which resembled a turtle shell. It was an individual choice as to which one you wanted to wear. Most of the regular flight crews opted for the lighter flak jackets in lieu of the forty-pound turtle shells. The turtle shell consisted of two thick pieces of composite honeycomb metal and fiberglass plate that were contained within a nylon vest. This gear was donned by slipping it over your head and fastening the sides with Velcro flaps. This afforded better protection against small arms fire and shrapnel, but made moving around very awkward. I carried both types of protection aboard my bird. I sat on the turtle shell and wore the flak jacket. I guess as a nineteen-year old, I valued my personal jewels more than my heart! On this particular mission Childers was wearing a turtle shell. The bullet hit him on the side, between the protective plates, and ricocheted through his chest. It tore him up real bad and he bleed to death in a few minutes. There was nothing Nixon could do for him except hold him in his arms, cry, and watch him slip away.

A door gunner's view of Vietnam from an HMM-263 Seahorse. (Photo courtesy of Leonard G. Martinez)

Maintenance on one of HMM-263's UH-34Ds at Marble Mountain in November, 1966. (Photo courtesy of Leonard G. Martinez)

The news reached me while I was standing in line for evening chow. By the time I found Nixon, he was in his hootch with a couple of the other flight crew guys. I got the details from someone else, since Nixon was too shook-up to talk. He just sat there sobbing, not able to speak. I didn't know what to say; I was totally lost for words. We were all having a hard time accepting the fact that Jim was dead. He was a real likable guy, always laughing and grabassing with us. I didn't really know him that well. We had flown a few missions together and shared a couple of beers at the E Club. He was much tighter with Nixon and a few of the other guys in the squadron. All that I can really remember about him was his big, grinning face, blond hair, and that he was from the San Francisco Bay area.

Later that evening someone brought us some beers. We proceeded to get good and drunk, swapping stories about some of the crazy things Jim had pulled off. In our minds we somehow had to rationalize that his death was a result of a mistake he had made. We didn't want to accept the fact that maybe it was just pure chance that got him killed, and we were just as vulnerable as he was. Looking back on it now, I guess it was our way of self-preservation; otherwise we would have had to accept the fact that surviving the war was due to luck and we really had no say in it. We didn't want to believe it could happen to us. I guess this has always been the way men have reacted in wars. You mentally bury your lost friends, closing them off for a later day. It's only years later when the hurt and loss really sets in. The next day the squadron rotated back to Okinawa and it was the last time I saw or heard from any of those guys again. I often wonder how many of them made it out safely.

Twenty years later, on the occasion of my first visit to Washington, D.C., I found my way to the Vietnam Memorial. I searched for Jim's name and found it near the top of one of the black granite panels, beyond the reach of my fingertips. It

happened to be raining that morning, which only intensified my memory of Nam and his loss. I remember standing there staring at his name, feeling the hurt of his loss as strongly as I had twenty years earlier. I didn't want to read any of the other names for fear that I would come across others buried in my past. I also thought about Nixon and the nightmare that he had to carry with him for the rest of his life.

Paul Gregoire, United States Marine Corps and Air America:

During my first Vietnam tour with HMM-163, in 1965, we operated two H-34 "Stinger" gunships for about three months. I believe this was the only combat flying that was done by that particular configuration. The bird was armed with two eighteen-rocket pods, one mounted on each side just above the landing gear a-frame. Two fixed M-60 machine guns were mounted to a plate affixed to each of the rocket pods. On the left side of the fuselage, a hole was cut into the skin, allowing ammo to be fed to those guns from inside the bird. As I recall, the ammo for the starboard guns was fed from a hole cut into the fuselage skin forward of the crew chief's position. It was a patchwork rig, but I enjoyed the chance to shoot back. The project was scratched because the H-34 was simply not suited to the mission: not fast enough, not maneuverable enough, and very prone to blade stall because of the weight. An Army Huey gunship platoon was assigned to our squadron for support.

I had two tours in Vietnam flying H-34s, and spent two-and-a-half years with Air America in Laos flying H-34s and TwinPacs (S-58Ts). One mission that comes to mind that specifically relates to the construction of the H-34 happened in September of 1967 in the Que Son Valley, southwest of Da Nang. Everyone who flew H-34s was scared to death of fire because of the magnesium skin. This tale shows why.

We were flying a routine resupply mission from the Logistical Supply Activity (LSA, otherwise known as a supply dump) at the mouth of the valley to various positions located further west in the valley. After doing a corkscrew climbout

"Trying to decide who's running the show." (Photo courtesy of Leonard G. Martinez)

Captain Paul Gregoire, USMC, mounts up at Ky Ha, north of Chu Lai, in 1967. (Photo courtesy of Paul Gregoire)

over the LSA to an altitude of 2,500 feet MSL (about 2,200 feet AGL), we headed west to our first destination. Approximately ten miles later we were hit by a burst of heavy automatic weapons fire, most likely from a 12.7mm antiaircraft machine gun. I felt the impact of several rounds hitting us and saw quite a few tracers going by the cockpit on both sides. Almost immediately the crew chief came up on the intercom and announced that we were on fire. He told me that he had emptied his fire bottle on the flames but that we had three separate fires going in the forward fuel cells.

I called my wingman and advised him that we were on fire and going down, then began an autorotation in an effort to get on the ground before the million-dollar flashbulb we were flying completely disintegrated. In just a few seconds I began to feel the heat rising up from the cabin, and at about that time the crew chief's smoke grenades began to self-ignite. The cockpit was soon filled with smoke in an assortment of colorful hues and we were unable to see anything. I opened my window fully and put the bird into a slight left crab to try to clear out some of the smoke. As I looked to my right and down I saw the crew chief standing on the landing gear A-frame outside the cabin! The fire had grown so intense that he had stepped outside and was hanging on to the strut for dear life. I found out later that the gunner had crawled into the electronics compartment in an attempt to get away from the fire.

Because of the smoke my forward vision was almost non-existent. When I estimated that my altitude was about 250 feet AGL I went into a side flare to clear the smoke. Much to my chagrin and surprise, instead of being at 250 feet we were at about 50 feet AGL. I immediately pulled all the collective I had and hauled back on the cyclic. It helped somewhat, but we still hit very hard in a side flare, tail first. The left main gear broke off when it hit, and I can remember seeing it going through the main rotor disk. The tail had also broken off on impact, and that was how the gunner made his successful escape. He exited the broken pylon carrying his M-60 machine gun. The crew chief was ejected from his perch when we made contact with the ground and was thrown clear. He wound up in a patch of brush about seventy-five feet away. To this day I swear I saw him go through the rotor disk when he was thrown clear, but his only injuries were a few lacerations.

After impact I reached up to apply the rotor brake. This was strictly habit as the rotor blades had hit the ground, broken off, and were definitely stopped. As I reached for the rotor brake I saw that it was already on. I glanced over to my co-pilot and saw that his seat was empty. It was then that I realized that the flames were already at seat level and the only thing between me and them was my Nomex flight suit. I looked to the right and saw the co-pilot standing outside my window waving and yelling at me. I dove out and landed on my knees in a dry rice paddy. My wingman was circling overhead and I called him with my survival radio. There was no place for him to land close to us so he directed us to a suitable LZ about two hundred meters away. Although we could hear Vietnamese voices behind us as we moved toward the LZ, we managed to keep them at bay with the two M-60s we had salvaged from the bird. We scrambled aboard the other helo about fifteen minutes later and made a safe departure.

My co-pilot and I both suffered back injuries from the "landing". I eventually was sent to Okinawa for therapy and rehabilitation. The incident provided my co-pilot with his third Purple Heart in two months and he was also sent to Okinawa. He later went back for a second tour in CH-53s. What was remarkable about the incident was that, as far as we could determine, no H-34 had ever started on fire at an altitude of two thousand feet or higher and made it to the ground with everyone getting out relatively unscathed. It was unusual enough that the Commanding General of the First Marine Air Wing came to see me and offer his congratulations.

As for recip versus turbine, there were lots of differences. Recips were much harder to fly well because of rpm control with the motorcycle-style throttle. The sign of a really good pilot was rock steady rpm control carrying very heavy loads, especially on takeoff and landing. Fluctuations of more than twenty apparent rotor rpm were the sign of a sloppy pilot. This was not a problem with the turbine because of the governor. All you had to do was beep it up if you needed more. It was automatic and much easier.

The turbine offered the great advantage of an extra engine. I had never thought much about losing an engine when I was flying single-engine aircraft, but I thought about it a lot when I started flying the TwinPac. Go figure. In about two thousand hours of TwinPac flying I only lost an engine once, and that was on takeoff from a paved runway with a light load. There was very little indication of power loss and we flew it back to homebase on one engine. The turbine also offered the great advantage of greater available power at high altitudes. This was extremely important to us in Laos, a very mountainous country.

The only advantage the recip had over the turbine was the instantaneous power response to throttle inputs. We used this ability extensively in Vietnam by making high speed, low level "button hook" approaches to hot landing zones. This type of approach is not possible with a turbine powered helicopter with its engine lag.

The turbine was less noisy than the recip and quite a bit smoother in the air. Both cockpits were basically the same—very uncomfortable and very tiring, even after only a few hours. With AAM we flew many fourteen to fifteen hour days, and I had a seventeen hour and forty-five minute day once. It's a good thing we were paid by the hour. Talk about a sore back!

I never had an engine problem in Vietnam that was not directly caused by ground fire. In the States in late 1968 I had a chip light come on over the Pacific while I was practicing a TACAN approach to El Toro. That subsequently became a rough runner followed by catastrophic engine failure just as we crossed the beach line. It only ran for about ten minutes after the chip light came on. Oil pressure went to zero, cylinder head temp pegged and I knew it was only a matter of time. We were pretty lucky, as there were no Mae Wests aboard.

When I was instructing at the Naval Air Training Command I flew T-28Bs and T-28Cs. Both had R-1820s, and it was very common to get chip lights. These were almost always caused by water in the sump, and were almost universally ignored barring confirmation from the gauges.

In Vietnam, it was common practice to do an engine change at seven hundred to eight hundred hours. Because of the dust, sand, rapid power changes, over-boosting, over-speeding, etc., the engines just did not last too long.

There is a vast difference in the handling characteristics of an internal load versus an external load. Experience tells me that the H-34 handles much better with an internal load even if it is heavier than an external load. The obvious reason is the shifting center of gravity (CG) caused by an external load swinging around due to wind or pilot-induced oscillations. As the load swings the CG shifts, causing the need for small control inputs. The normal tendency of inexperienced pilots is to overcontrol when an external load starts to swing. With a heavy load on a long line this tendency can soon cause catastrophic control problems. Very heavy loads require an experienced pilot.

Carrying heavy loads (like Air America did by strapping a wrecked Pilatus Porter to the fuselage) would require that the H-34 be stripped of all extraneous weight. This would include, but not be limited to, all tools, spare oil, blade boots, etc.; all unnecessary radio and navigation equipment; all doors and cargo compartment seats and steps; all armor plate including the laminated armor under the clutch compartment; carrying only enough fuel to complete the mission with a minimal reserve; and limiting the crew to one pilot and one crew chief. This is not a routine mission for an H-34.

I recall an incident in Vietnam where we were supposed to carry live cattle externally. The poor beasts were blindfolded but not drugged in any way. As the first bird came to a hover over the terrified animal, it was almost impossible for the hookup crew to control it. Finally the connection was made and they lifted off. As soon as the cow left the ground she started to violently kick her legs in a running motion, causing her to swing back and forth in a wide arc. It was obvious from the ground that the cow was causing horrible control problems for the pilot. At about five hundred feet he pickled her off before he lost control. Needless to say, that was one dead cow.

Internal loads are generally stable and don't cause the CG problems that externals do, as long as they are loaded correctly. I don't think that the external 20mm turret tried by the French would cause any particular problems other than those caused by any internal load of corresponding weight. The major problem I see would be the same one we had with the "Stinger". The more the bird weighs, the lower the retreating blade tip stall speed. Attacks on ground targets required a dive to put ordnance on the target, either with the rockets or the guns. As heavy as we were with thirty-six rockets, six M-60 machine guns and thousands of rounds of ammo, smoke grenades, etc., our blade stall speed was around 110 knots or even less. It didn't take much of a dive to exceed that speed. Almost every attack we made was right at the very edge of blade stall.

I've got another "sea story" about CG control, this time with an internal load. In '65 I was co-pilot on a mission where we were moving a village population to a controlled area. We were hauling everything they owned, including all their animals. We had a load of men, women, children and assorted chickens and pigs. About half way to our destination the aircraft suddenly went into a nose-high attitude. The aircraft commander shoved the stick forward to get the nose down but was only partly successful. With the cyclic fully forward and the droop stops thumping away, he called to the crew chief on the ICS to find out what was happening. The crew chief answered that one of the pigs had gotten loose and run into the electronics compartment, and his owner had followed to catch him. He asked if we wanted him to go back and get the guy! He was told emphatically to stay where he was and to pull some of the people near him farther forward. A few seconds later the pig and owner came out of the electronics compartment and we regained control. If our two hundred-plus pound crew chief had also gone back to the rear, I'm sure our flight would have ended right there.

On the subject of vibrations, a "one-to-one" was caused by a blade being out of track, causing one "bump" per rotor revolution. Frequently, you could see one blade "jumping" from the cockpit. The problem was easily fixed by an adjustment to the rotor head. When I was a junior Second Lieutenant I frequently was given the job of blade tracking. It was a very boring procedure for the pilot, which consisted of engaging the rotor and turning it at varying rpms while the maintenance folks adjusted the blades. Each blade tip was marked with a different color, then a canvas tracking "flag" was eased into the rotor disk while the blades were turning. The position of the various colors on the flag would indicate which blades were not tracking properly.

Vibrations were very common, and not a cause for concern unless it was a high frequency vibration. You could feel a "high freq" by a tingling in your extremities, such as your hands, feet, and especially your nose. It was an indication of a potentially serious drive train/rotor condition, and called for grounding the bird for maintenance checks as soon as possible.

Dave Bjork, United States Marine Corps:
The only H-34 time I had was in the training command at Ellyson Field just outside of Pensacola, Florida. With that large radial engine, the H-34 was really a noisy helo. I wasn't looking forward to my first night navigation flight. We took off about ten at night. In Florida, in August, it was really hot and muggy. We started the route out over some desolate country. The instructor opened his window for a little air, and the noise was deafening. After about an hour, I was really concentrating on flying and navigating. All of a sudden, the instructor slammed his window. I didn't see him do it, so the first thing I thought was that the engine had failed and we were going to autorotate down into the Florida swamps. For about five seconds I thought I was surely going to die.

Pat Kenny, United States Marine Corps:
I had thirty-four hops that gave me a grand total of sixty-seven hours in the H-34 while I was at Pensacola with HT-18 in August and September of 1969. I have to say that I was very glad to get the time, as I believe I was in one of the last (if not the last) classes to do so. Others at HT-18 were transitioning during the same period into the UH-1. It prepared me for the sling work that we would do later in the CH-46, flying out of Marble Mountain. I flew in the SH-34J, SH-34G, UH-34D, UH-34G, and the UH-34J.

The memory I have that remains forever etched in my head is the checklist on the H-34. The checklist was on a hinged Plexiglas board that rested on the top of the instrument panel and flipped down so one could follow it. I remember the first time I encountered it, I was a bit overwhelmed. It must have been at least two-and-a-half feet across and a foot high. I haven't experienced anything like it since.

My final check ride was done along with three other students on an all-night excursion with a Captain Campbell. He was an instructor who had been working at Sikorsky when he reactivated and came back into the active Corps as a flight instructor at HT-18. He was older, very patient, and wise. He knew the H-34 very well and was a character to boot. That last flight lasted about fourteen hours, as each of us flew several hours of First Pilot time to complete the syllabus. I think they were in a hurry to meet the quota for the meat factory in Southeast Asia. It was a clear night, and we flew from Pensacola down along the coast to New Orleans and then north to God knows where and made this huge circle. I remember seeing the sun come up on the horizon (we had started out at about dusk the evening before) and thinking to myself that Captain Campbell must have been a very tired instructor, because this flight was non-stop except for refueling and a sandwich stop as we went.

Bob Steinbrunn, Imperial Helicopters:

I thought the Sikorsky S-58 was built like a bridge: strong, massive structural members, imposing, and not easily damaged. Despite its size, the civilian versions I flew were light and were very maneuverable and responsive. With its dual hydraulic systems pressurizing the main servos which activate the flight controls, it was very light on the controls, took no effort to move the cyclic or collective (unlike the bus-like S-55 which took real muscle to fly), and was a delight to fly.

Starting the Wright Cyclone R-1820-84C engine was an art requiring the right amount of prime and throttle. The juggling of mixture control, throttle, primer, and magneto switch during the initial coughing of the engine in order to get it to settle down into a steady, throaty roar took time and experience on the part of the pilot. It was a source of satisfaction to get one lit off on the first try.

Unlike turbine engines, reciprocating engines are very difficult to start in cold weather. Just as in a car, the battery has less cranking power when it's cold, so some form of external power cart is usually used. Next, the fuel atomizes less well in the carburetor as the temperature grows colder and engine light-off becomes more difficult. Some form of engine pre-heating is recommended when the temperature is below freezing.

Lastly, and most importantly, the engine lubricating oil thickens and can actually congeal in the engine and oil coolers. The oil used in the civilian S-58s I flew was Aeroshell 100, which was a 50-weight (!!) non-detergent, thick, heavy-duty oil which made it very difficult to turn the engine over in cold weather. The oil, when cold, became like glue and gripped the pistons, rings, connecting rods, and other major engine parts, and could actually prevent turning over the engine with the starter. If the starter could turn the engine over, you ran the risk of burning it out. And if the engine did manage to light off, the oil could be too thick to properly lubricate the engine, and scuffing of pistons and cylinder walls could occur. This is not a happy situation. Obviously, if some means could be found of artificially thinning the oil temporarily, the engine could be easily started in cold weather.

Enter the Oil Dilution System. With either automatic or manual control, the pilot could, on shutting down a warm engine after a flight, activate a switch which introduced AvGas from the fuel tanks into the engine oil system, effectively thinning it down which would allow easy engine starting for the next flight. When the engine was next started, and as it warmed up, the heat from engine operation vaporized the fuel in the oil, the fumes from which exited via the crankcase breather, and the oil then returned to its original viscosity for proper lubrication of the hot engine.

In moderately cold temperatures, using oil dilution in "automatic" was sufficient, the system metering in a certain amount of fuel at a prescribed rate.

In unusually cold temperatures, the pilot could select "manual" in order to manually meter in more AvGas to thin the oil much more than the "automatic" setting would allow.

As you may see from the above, cold starting was a hot topic (ouch).

The "round motor" versions I flew were very noisy inside since all of the insulation was stripped out of them in an effort to make them lighter. They were, after all, heavy lifters. Only the David Clark sound-suppressing headset was of high enough quality to protect your hearing.

The S-58Ts I flew were converted with the Pratt & Whitney PT6T-3 TwinPac engine. Essentially two PT-6s mounted side by side and shafted into a combining gearbox, this engine installation made the aircraft much quieter inside as well as much more vibration free. The rotor head had Sikorsky's Bifilar vibration absorbing device installed which also helped to make the airframe very smooth. The "T" was even more of a delight to fly since it had single-engine flight capability, provided you were light, and had more current systems, controls, instrumentation, and avionics than the older piston engine models.

Much like putting turbine engines on a DC-3, the S-58T was a case of mating a new-technology engine with an old-technology airframe. It made the old girl a wonder, but her time had passed. She was truly the DC-3 of the helicopter fleet.

Chapter 3
Argentina

Navy

The *Escuadrilla Aeronaval de Helicópteros* (Naval Helicopter Squadron) operated the Argentine Navy's first helicopter obtained and equipped specifically for ASW duties, as well as its only H-34. This HSS-1 was delivered 17 October 1957 as s/n 0407, with the code 2-HT-10, and recoded 2-HP-1 before being officially placed in service on 19 December of the following year. Although intended for ASW, 0407 also participated in a number of rescue flights, including one that resulted in its crash on 5 June 1959 while coded 2-HP-406. After repairs, its code was changed to 2-H-10 in 1960. A preflight fire destroyed 0407, by then coded 2-HT-21, on 25 April 1961.

The sole Argentine Navy Seabat, 0407 (c/n 58-611), hovers at the Bridgeport plant in October, 1957. The gear-mounted tail float installation is unusual, as is the "Sikorsky S-58" legend under the cockpit. (Photo S22901D courtesy of Sikorsky Aircraft Corporation)

Argentine Air Force S-58ET H-02 (c/n 58-740) between 1975 and 1977. (Photo by Jorge Nuñez Padin courtesy of the author)

Air Force

The *Fuerza Aérea Argentina* (Argentine Air Force) operated more H-34s than did the Navy, but service was even shorter than that of the single HSS-1. Two S-58Ts, converted from military airframes by Carson Helicopter of Pennsylvania, performed Presidential transport and SAR duties from 1975 until 1977. They were coded H-01 and H-02. Both aircraft went to the Argentine civil registry after Air Force service.

During the conflict over the Falkland Islands, civil aircraft were pressed into service by the *Fuerza Aérea*. Among these were the original two Air Force S-58Ts. Now jointly operated by Court Helicopters (of South Africa) and *Helicópteros Marinos S.A.* (of Argentina), they were registered LV-OCM and LV-OCN. They served between 18 May and 23 June 1982 in SAR and VIP transport roles. LV-OCN reportedly tangled with a powerline, but was not destroyed in the accident. Both aircraft returned to civil service after their period of impressment, and were later sold to US owners.

model	c/n	s/n	notes
HSS-1	58-611	0407	with various codes
S-58ET	58-740	H-02	ex-H-34G.I, impressed while LV-OCM
S-58DT	58-1464	H-01	ex-HUS-1 149362, impressed while LV-OCN

H-02 became LV-OCM in Helicopteros Marinos service. Impressed during the Falklands War, it was damaged when it tangled with a powerline. The color scheme is that of Court Helicopters, with whom Helicopteros Marinos operated several ex-military S-58s. (Photo by Ken Hudson courtesy of the author)

Chapter 4
Belgium

The Belgian Air force (known in Flemish as the *Belgische Luctmacht* or LuM, and in French as the *Force Aérienne Belge* or FAéB) formed its Search and Rescue Flight around Sud Est-built HSS-1s at Koksidje on 11 April 1961. The original two aircraft, OT-ZKD and OT-ZKE, had been delivered six days earlier, following a series of familiarization and training flights with the French *Aéronavale's Flotille 23S* at Saint Mandrier. OT-ZKF joined the unit on 12 January 1962, with OT-ZKG and OT-ZKH being received on 21 February and 20 April of that year. These last two aircraft were transferred to the *Force Navale* (Belgian Navy) in 1974, but still operated from Koksidje. ASW gear was never fitted; instead, troop seating for sixteen was installed in the main cabin.

The Short Range Transport (SRT) Flight operated the only European military S-58Cs beginning in 1963. Five were acquired from Sabena, which had used them successfully on inter-city passenger routes since late 1956. OT-ZKK and OT-ZKL were taken on strength 8 May, with OT-ZKJ coming on the 28th of that month. OT-ZKI and OT-ZKM were added on 21 October. The sixth and seventh examples came from Sabena as well, having originally been leased by Sabena from Chicago Helicopter Airways in 1963. These were OT-ZKP, obtained on 7 January 1969, and OT-ZKN on 12 June. On the first five, rescue hoists were fitted, some of the fuselage glazing was covered over, and the plush interior was replaced with military webbed seats. This was not the case on the last two, which served as VIP transports and retained all of their commercial furnishings.

The SAR and SRT Flights were joined under the *Escadrille Heli* (Heli-Flight) on 1 April 1971. Also tasked with mine clearing, surveillance, and transport roles, this became *40éme Escadrille Heli* on 30 October 1974. Ten of the twelve Sikorskys were still in service at this time; OT-ZKI was written off following a June, 1964, crash, as was OT-ZKM on 15 October 1971. The remaining five S-58Cs were all withdrawn from service by the end of 1976, and eventually sold on the German civil market. Of the HSS-1s, OT-ZKF was retired on 10 May 1979, OT-ZKD on 5 December 1984, and OT-ZKE was written off on 29 August 1985. Having been transferred to the Belgian Navy's Heli-Flight, OT-ZKG crashed on 7 January 1976, while OT-ZKH made the final flight of a European military H-34 on 19 July 1986. Along with being the last active example on the Continent, it was the last Sud Est-built HSS-1, and it is fitting that it was painted in 40 Squadron colors with special markings for the final flight. The HSS-1s had, by this time, amassed from just under three thousand to well over forty-five hundred hours of flight time. Including their earlier civil service, almost all of the S-58Cs had clocked over eleven thousand hours per airframe.

Although the Belgian HSS-1s served well past the point where examples in most other militaries became SH-34Gs, and despite the fact that being void of ASW gear qualified

HSS-1 OT-ZKD/B4 (SA-145). (Photo courtesy of Ronald W. Harrison)

HSS-1 OT-ZKE/B5 (SA-146) on 12 June 1969. (Photo courtesy of CFAP)

A starboard view of HSS-1 OT-ZKH/B8 (SA-185) in its special retirement scheme. (Photo M06548 courtesy of MAP)

OT-ZKH's retirement scheme from the port side. (Photo M19579 courtesy of MAP)

S-58C OT-ZKI/B9 (c/n 58-324) originally flew with Sabena as OO-SHG. It also flew briefly in Katanga as KAT43. (Photo courtesy of Rudy Binnemans)

Received on 21 October 1963, OT-ZKI was stricken eight months later, following this June, 1964, crash. (Photo courtesy of Rudy Binnemans)

S-58C OT-ZKK/B11 (c/n 58-356) at Florennes, Belgium, in June, 1975. It was once Sabena's OO-SHI. (Photo by Richter courtesy of Terry Love)

S-58C OT-ZKN/B14 (c/n 58-350) saw prior service with New York Helicopter and Chicago Helicopter as N878, and with Sabena as OO-SHP. Unlike most of the S-58Cs, it retained its plush interior during military service. (Photo courtesy of Ronald W. Harrison)

them to be renamed UH-34Gs, the LuM followed the French example of retaining the old designation until the type was retired. Not being comparable to any model built for military service, the S-58Cs retained their civil designation.

model	c/n	s/n	notes
HSS-1	SA-145	OT-ZKD/B4	
HSS-1	SA-146	OT-ZKE/B5	
HSS-1	SA-181	OT-ZKF/B6	
HSS-1	SA-184	OT-ZKG/B7	crashed
HSS-1	SA-185	OT-ZKH/B8	
S-58C	58-324	OT-ZKI/B9	ex-Sabena OO-SHG; crashed
S-58C	58-333	OT-ZKJ/B10	ex-Sabena OO-SHH
S-58C	58-350	OT-ZKN/B14	ex-New York Airways, Chicago Helicopter Airways N878; ex-Sabena OO-SHP
S-58C	58-356	OT-ZKK/B11	ex-Sabena OO-SHI

Parts of OT-ZKM/B13 (c/n 58-395) following its destruction in a crash on 15 October 1971. This S-58C served earlier with Sabena as OO-SHM, and in Katanga as KAT44. (Photo courtesy of Ben Marselis)

S-58C	58-388	OT-ZKL/B12	ex-Sabena OO-SHL
S-58C	58-395	OT-ZKM/B13	ex-Sabena OO-SHM; crashed
S-58C	58-836	OT-ZKP/B15	ex-CHA N869, ex-Sabena OO-SHQ

LEFT: A pair of electrically triggered flare dispensers was mounted on the starboard oleo of Belgian HSS-1s. (Photo courtesy of Rudy Binnemans)

HSS-1 OT-ZKG/B7 (SA-184) during SAR training after transfer to the Belgian Navy. (Photo courtesy of Rudy Binnemans)

Chapter 5
Brazil

Air Force

The *Força Aérea Brasileira* (FAB, Brazilian Air Force) obtained six HSS-1Ns under the Mutual Defense and Assistance Program (MDAP). Training using them took place at NAS Key West. The aircraft were then shipped to Rio de Janeiro in pairs aboard USAF C-124 Globemasters (the only aircraft in US service at the time that could carry H-34s). The first two, FAB serials 8551 and 8552, arrived at the *2°. Esquadrão do*

FAB HSS-1N 8552 (c/n 58-1295) later became Brazilian Navy SH-34J 3003. (Photo by Francisco C. Pereira Netto courtesy of Sergio L. dos Santos)

After FAB service, HSS-1N 8555 (c/n 58-1309) served in the Marinha as SH-34J 3006. (Photo by Francisco C. Pereira Netto Sergio courtesy of L. dos Santos)

Marinha 3003, formerly FAB 8552, in the original Naval Aviation scheme. (Photo courtesy of the Brazilian Navy)

FAB 8555 became Marinha 3006. (Photo by Francisco C. Pereira Netto courtesy of Sergio L. dos Santos)

Marinha SH-34J 3001 (c/n 58-1283) in the second Naval Aviation color scheme. (Photo courtesy of the Brazilian Navy)

1°. Grupo de Aviação Embarcada (Second Squadron of the First Group of Embarked Aviation) on 19 February 1961; 8550 and 8553 followed on 2 March, 8554 and 8555 on 4 April. Under the regulations of the time, the FAB operated these Seabats in the training, transport, and SAR roles until 1965.

Navy

As a result of Presidential Decree 55.627 of 26 January 1965, the FAB Seabats were transferred to the *Marinha do Brasil* (MB, Brazilian Navy) and re-serialed N-3001 through N-3006. This transfer resulted in the creation of *1°. Esquadrão de Helicópteros Anti-Submarino* (HS-1, First Anti-Submarine Squadron) by Ministerial Advice 830 on 28 May 1965. The aircraft were actually taken on strength by the MB between 29 June 1965 (N-3004 and 3006) and mid-1966 (N-3005). HS-1 regularly operated from the carrier MINAS GERAIS until August, 1974. During that time, N-3003 was destroyed in a crash at the Sao Pedro da Aldeia Naval Air Station on 8 Feb-

ruary 1966, and N-3005 crashed at Mount Sao Luiz on 10 November 1972. The Director of Naval Aviation ordered the remaining four SH-34Js decommissioned in August, 1974. The airframes were scrapped, and spare parts became property of the Ministry of Mines and Energy.

Model	c/n	FAB	MB	MDAP
HSS-1N	58-1283	8550	N-3001	60-5424
HSS-1N	58-1294	8551	N-3002	60-5425
HSS-1N	58-1295	8552	N-3003	60-5426
HSS-1N	58-1307	8553	N-3004	60-5427
HSS-1N	58-1308	8554	N-3005	60-5428
HSS-1N	58-1309	8555	N-3006	60-5429

Some sources show BuNos 148934 through 148939 (c/ns 58-1301 and 58-1310 through 58-1314) for these aircraft. However, two of those serials were assigned to SH-34Js which survived into the US civil market, which the Brazilian aircraft clearly did not.

Like the other Brazilian H-34s, Marinha 3004 (c/n 58-1307) was eventually broken up for parts and scrap. (Photo by Francisco C. Pereira Netto courtesy of Sergio L. dos Santos)

Chapter 6
Canada

Canada was the second foreign purchaser of H-34s (after France), operating six over a seventeen-year period. The first three H-34As (9630-9632) were delivered on 3 November 1955, and the remainder (9633-9635) on 12 January of the following year. The provided yeoman service during the mid-1950s in the construction of the early warning defense system known as the Mid-Canada Line. Operating units included 448 Squadron at Cold Lake; 5(Hel) OTU and 108 Communications Flight at Rockcliffe; 102 Helicopter Flight, 102 KU, and 4(T) OTU at Trenton; and 111 Communications Unit at Winnipeg. Over the course of these varied assignments, their duties included SAR, cargo and personnel transportation, and training. 9631 was fitted with steps, a new cabin door, and plush interior for its role as a VIP transport, which included Princess Margaret's 1958 Canadian tour. On one occasion, an RCAF Choctaw was instrumental in tracking bank robbers and assisting in their capture.

Sikorsky records show these aircraft as S-58Bs. A review of RCAF manuals, however, indicates that all equipment was to H-34A standards. The switch in designations is not surprising or unusual, as company records for the earlier S-55 list virtually every exported airframe under a civil designation. Although Canadian H-34As were "stock" in all respects, unlike some foreign examples, at least one was modified while in service to carry an F-86 drop tank on the port side of the fuselage. This was different from factory installations for the US military.

9634 was written off following a crash on 30 November 1958 while with 111 Communications Unit, and 9631 suffered the same fate with the same unit on 17 April 1961. 9630 was declared surplus in the summer of 1966. The surviving three (9632, 9633, and 9635), were retired out of CFB Cold Lake between 5 November 1971 and 26 January 1972, with a total of over 15,000 hours between them.

9630 (c/n 58-160) was the RCAF's first H-34A. (Photo courtesy of Canadian Forces Photo Unit)

model	c/n	s/n	notes
H-34A	58-160	9630	to US civil
H-34A	58-161	9631	crashed 26 June 1961
H-34A	58-202	9632	to Canadian and US civil
H-34A	58-223	9633	to Canadian civil
H-34A	58-224	9634	crashed 10 February 1959
H-34A	58-245	9635	to Canadian and US civil

Just visible under the fuselage of 9631 (c/n 58-161) are the special cabin steps added when it was converted for VIP duties in 1958. (Photo courtesy of Canadian Forces Photo Unit)

The last Canadian H-34A, 9635 (c/n 58-245), carries supplies during construction of the Mid-Canada Line. Sikorsky assigned c/n 58-245n to HSS-1 141591, a potential source of confusion when tracking identities. (Photo courtesy of Canadian Forces Photo Unit)

Chapter 7
Chile

The Chilean Navy acquired two Seabats from the Sikorsky production line under the Military Assistance Program (MAP), although the purchase of four had at one time been planned. N-51 was delivered on 9 May 1962, and made its first flight on 6 September. N-52 arrived on 18 April 1963. They operated with the *Escuadrón Antisubmarino* (Antisubmarine Squadron) as ASW and SAR aircraft, in concert with ships of the *Armada de Chile*, but because of their size were shorebased.

Retirement came in 1978. N-51 became a museum piece at NAS Vina del Mar. Its companion was used for a period as a ground trainer for paratroops, then sold to a scrap dealer who chose to restore it for the *Museo Nacional de Aeronáutica*.

model	c/n	s/n	MAP s/n
HSS-1N	58-1429	N-51	BuNo 149840
SH-34J	58-1623	N-52	BuNo 150730

These have been erroneously reported as three aircraft, with BuNos 150730-150732 and Chilean serials 51-53.

LEFT: HSS-1N 51 (c/n 58-1429, BuNo 149840) during a routine sortie off the coast of Chile. (Photo courtesy of Museo Nacional de Aeronáutica de Chile)

RIGHT: Chilean SH-34J 52 (c/n 58-1623, BuNo 150730). Seabats were too large to fly from Chilean ships, and so were limited to coastal patrols. (Photo courtesy of Museo Nacional de Aeronáutica de Chile)

Chapter 8
Costa Rica

While it has no armed forces *per se*, Costa Rica's defense needs are met by the *Ministerio de Seguridad Popular* (MSP, Security Ministry), sometimes referred to as the *Guardia Civil*. The few aircraft flown by this organization are carried on the civil aircraft register in the TI-SPx series.

Two ex-USAF HH-34Js, themselves ex-USN SH-34Js, have been operated by the MSP. These were taken from storage at Davis-Monthan AFB and first appeared on the Costa Rican register in October of 1975. Almost no other details are available. TI-SPI was derelict at San Jose by February of 1980, and might have become a museum piece in Costa Rica. There is no information to suggest that either survived to the civilian market.

model	c/n	s/n	notes
SH-34J	58-1050	TI-SPI	ex-BuNo 145707
SH-34J	58-1353	TI-SPJ	ex-BuNo 148957

TI-SPI has also been erroneously reported as TI-SPF, and as BuNo 145704.

LEFT: TI-SPI (c/n 58-1050, ex-SH-34J 145707) lies abandoned at San Jose, Costa Rica, in February, 1980. (Photo by Dr. Gary Kuhn courtesy of Dan Hagedorn)

RIGHT: Also at San Jose, here in January, 1980, TI-SPJ (c/n 58-1353, ex-SH-34J 148957) later became a museum display. (Photo by Dr. Gary Kuhn courtesy of Dan Hagedorn)

Chapter 9
England

While the Royal Navy and Royal Air Force operated large quantities of the Westland Wessex, only one Seabat airframe found its way to military service in England. This was HSS-1 XL722, purchased by Westland as the prototype for the Wessex. Delivered on 15 May 1956 with a Cyclone engine, it was re-equipped with a Napier Gazelle NGa 11 gas turbine and first flew in this configuration one year and two days later. After three years of service at Yeovil as an engine and development testbed, it saw use at RNAS Arbroath between 1960 and 1964 as a Fleet Air Arm instruc-tional airframe. Royal Navy maintenance serial A2514 was assigned in the second half of 1961. It was in a scrapyard at Ascot by early 1964, and by 1968 had been destroyed for scrap.

model	c/n	s/n	notes
HSS-1	58-265	XL722	originally G17-1 (Westland)

This airframe was to have been HSS-1 BuNo 141602; the number was reassigned to HSS-1 c/n 58-444.

Soon after delivery, HSS-1 XL722 (c/n 58-265) had its Cyclone replaced with a Napier Gazelle Nga 11 gas turbine. (Photo M11364 courtesy of Westland Helicopters)

XL722 with its second turbine engine installation, a more powerful NGa 13 that required greater changes to the nose. (Photo M13371 courtesy of Westland Helicopters)

Four of XL722's Wessex offspring displayed at the International Helicopter Museum in 1996. From left to right: civilian 60 Series G-AVNE, utility HU.5 XT472, anti-submarine HAS Mk.3 XS149, and HAS Mk.1 XM330. (Photo courtesy of Geoff Russell)

Chapter 10
France

Air Force

The *Armée de l'Air* (AdA, French Air Force) was Sikorsky's first foreign customer for the H-34, and the largest operator outside the United States. The first three Sikorsky production aircraft were delivered in 1956, after US military examples were already in service. By the delivery of SA-180 in 1962, the AdA took a total of 215 H-34s on strength. Production included seventy-nine completed by Sikorsky, Sud Est assembly of thirty-eight from Sikorsky-produced components, and the licensed building of ninety-eight by Sud Est. These sources produced aircraft that were essentially identical, differing only in changes or adaptations made to comply with the metric system and French radio requirements.

The impetus for this fleet of H-34s was the war being waged in Algeria. Here, the Air Force was responsible for troop transport, medevac, supply, and SAR west of the fourth degree of longitude (although operations frequently spilled over into the Army's area of responsibility east of that line). *Groupe Mixte d'Hélicoptères* (GMH, Composite Helicopter Group) 57 began operations from Boufarik, south of Algiers,

in June of 1956. Between then and the end of the war, the H-34s performed yeoman duty from bases at Boufarik, La Senia (near both Oran and the Moroccan border) and la Reghaïa. *Escadre d'Hélicoptères* (EH, Helicopter Squadron) 2 and EH 3 formed on 1 November 1956, the latter transitioning to H-34s after operating Bell 47Gs, Sud SE-3130 Alouette IIs, and Sikorsky H-19Ds. These units became EH 22 and EH 23 on 23 March 1961.

Dubbed *Mammouth* (Mammoth) by the French, unarmed H-34As undertook missions of up to three days' duration. Although built with sixteen-troop seating, the practical load in the often-mountainous region frequented by the H-34s was only twelve. Seats were commonly folded, or removed entirely, to ease exiting the aircraft in an LZ.

While there were no official programs for the installation of armor or armaments in helicopters in Algeria, the need for both was recognized quickly. French-designed bucket seats in the cockpit were fitted with fiberglass panels in the pan, and steel plate was shaped to cover the exposed oil cooler in the lower nose.

An American in Paris: the second Sud Est-assembled H-34A (c/n 58-455). (Photo 779 courtesy of AAHS and Maurice Salbert Collections)

Last of the component HSS-1s, SA-50 (c/n 58-1007) was used in 1969 to test Super Frelon sponsons. (Photo B87-1601 courtesy of SHAA)

H-34A SA-33 (c/n 58-755) carries a typical Algerian service weapons mix. A .50-caliber HMG can be seen on either side of the cabin, and an MG151 20mm cannon is in the door. (Photo B83-4236 courtesy of SHAA)

EH 2 flew the first armed helicopters in Algeria. The first of *lés Pirates*, as the armed H-34s were known, carried an incredible collection of 20mm cannon, .30- and .50-caliber machine guns, 73mm rockets, and bazookas of the same size. This was winnowed down to the 20mm and .50-caliber, although triple .30-caliber guns could be fitted on a twin-fifty mount.

Following the war, EH 22 moved from Oran to Chambéry, with detachments in Pau, Istres, and Cazaux. EH 23 went from la Reghaïa to Saint Dizier, and fielded detachments in Vallacoublay and Lahr (West Germany). In August of 1964, EH 22 and EH 23 came under the umbrella of the new

Commandement des Transports Aériens Militaires (COTAM, Transport Command). EH 22 became EH 1/68 (Pau, with a detachment at Cazaux), and EH 2/68 (Chambéry, with a detachment as Istres). EH 23 became EH 1/67 (Bremgarten, West Germany), EH 2/67 (Saint Dizier), and EH 3/67 (Villacoublay). By the early 1070s, EH 1/67 had moved to Cazaux, EH 2/67 to Metz, and EH 4/67 had been formed at Apt to work with the Strategic Missile Air Force Base there.

Additionally, AdA H-34As served with GMHs 67 and 68, and *Escadrille de Liaisons Aériennes* (ELA, Communications Squadron) 44, throughout Metropolitan France, as well as with the helicopter pilot training center at Toulouse-Francazal

HSS-1 SA-40 (c/n 58-932) is equipped with rearview mirrors and a mine sled towbar, an arrangement also seen on some American and German Seabats. (Photo courtesy of Frederick G. Freeman)

H-34A 58-1122 of GMH57 leaps into the air. Full nose armor is carried, but no port-side weapons. (Photo B87-2528 courtesy of SHAA)

The last Sikorsky-built HSS-1 delivered to France was 58-1376, photographed on display 12 September 1992. Despite its late c/n, it has the early pattern landing gear and kicksteps. (Photo by M. Fournier courtesy of Bob Burns)

H-34A SA-98 of EH 2/68 in medevac colors. (Photo Col. 865 courtesy of Ronald W. Harrison)

Section Jeanne d'Arc HSS-1s SA-136, SA-125, SA-143, and 58-550 while the ship was docked in Japan. (Photo by Yoshitaka Kato courtesy of Terry Love)

HSS-1 SA-150 with rescue basket and hub-mounted floats. (Photo M00596 courtesy of MAP)

EH 2/68 H-34A SA-154 retains its nose armor after the fighting in Algeria had ended. The external power receptacle was kept in its original position, aft of the cabin, on French aircraft. (Photo courtesy of Terry Love)

Seen abandoned in Chapter One, H-34A SA-168 in better days with EH 2/67. (Photo courtesy of Ronald W. Harrison)

A full assortment of weapons on a Pirate H-34A in Algeria: 23 air-to-surface rockets and bazooka launchers, one .30-caliber MG, two .50-caliber HMGs, and an MG151 20mm cannon. (Photo B87-675 courtesy of SHAA via Albert Grandolini)

An experimental 20mm cannon mount on an unidentified H-34A. (Photo B87-162 courtesy of SHAA via Albert Grandolini)

(later at Chambéry). Some also served with French forces in Chad, Djibouti, and Libya as elements of *Groupe Aérien Mixte d'Outre-Mer* (Overseas Composite Aviation Group) 00/088. Along with standard SAR, transport, and liaison duties, H-34As were used for VIP flights (including President de Gaulle) and ceremonial missions during the 1968 Grenoble Winter Olympics. The last of the type were retired from ELA 44 in 1974.

Navy

The *Aéronavale* (AN, French Naval Aviation) took a total of fifty-eight HSS-1s on strength. Fourteen were completed by

HSS-1 SA-127 was one of the testbeds used by *Escadrille* 20. On the starboard side were a Nord wire-guided missile and what might have been a napalm tank; there is no visible plumbing to suggest the object is a fuel tank. (Photo B87-1603 courtesy of SHAA)

Sikorsky, Sud Est assembled twelve Sikorsky airframes, and thirty-two were built under license by Sud Est. As with French Air Force H-34As, these sources produced aircraft with few differences.

Flottille 32F took its first deliveries in January of 1958, and 33F began receiving its in September of the next year. These units, along with 31F, which operated only a few of the HSS-1s, were part of *Groupement Hélicoptères Aéronavale* (GHAN) 1. In Algeria, the Group operated primarily from bases at Setif, Sidi Bel Abbes and Lartigue. Although not as heavily armed as their AdA counterparts, some of the GHAN 1 HSS-1s were fitted with a door-mounted 20mm MG151 cannon to suppress groundfire. 31F left Algeria in September of 1961, 32F on 26 July 1962, and 33F in August.

On their return to France, the HSS-1s of 31F and 32F were converted for ASW duties by the installation of French-built Alcatel DUAV-1 sonar. 33F continued in the troop transport role, serving most notably with French forces in Chad where it rotated with *Groupe Mixte de Transport* (GMT, Composite Transport Wing) 59 of the *Armée de l'Air*. There was a *Section Jeanne d'Arc*, attached to the carrier of that name, which provided third-year midshipmen at the French Naval Academy with an around-the-world training cruise. There was also a test and experimentation *Escadrille*, designated 20.S, at Saint Raphaël. With a total of 185,800 hours of flight time behind them, the last French HSS-1s were retired from GHAN 1 on 22 June 1979.

Armée de l'Air (215 aircraft)
79 complete H-34As were purchased from Sikorsky; c/ns were 58-: 1-3, 248, 266-267, 280-281, 299-300, 319-320, 328-331, 334-337, 340, 354-355, 357-358, 370, 374-379, 397-401, 421-422, 447, 478, 481, 488, 520, 553-555, 584-586, 593-595, 601-603, 610, 635-636, 659, 956, 977, 1046, 1070-1071, 1117-1130

38 H-34A airframes were delivered as components, assembled by Sud Est, and assigned their serials SA-1 through SA-38; c/ns were 58-: 435, 455, 469, 479, 487, 494, 500, 507, 513, 521, 525, 533, 541, 549, 556, 561, 567, 576, 582, 587, 609, 615, 637, 643, 653, 670, 682, 693, 705, 715, 729, 741, 755, 766, 781, 793, 819, 832

98 H-34As were built under license by Sud Est; c/ns were SA-: 51-118, 151-180

Aéronavale (58 aircraft)
14 complete HSS-1s were purchased from Sikorsky; c/ns were 58-: 445, 453-454, 512, 550, 638-641, 680-681, 688-689, 1376

12 HSS-1 airframes were delivered as components, assembled by Sud Est, and assigned their serials SA-39 through SA-50; c/ns were 58-: 922, 932, 944, 954, 961, 971, 983, 994, 1004-1007

32 HSS-1s were built under license by Sud Est; c/ns were SA-: 119-144, 147-150, 182-183

The Sud Est production numbers above do not include aircraft for Belgium.

Also reported in French service, but unconfirmed and in conflict with Sikorsky allocation lists, are c/ns 58-53 to 54, 74, 88 and 89, 318 and 717 for the AdA; and c/ns 58-1001 through 1003 for the AN. All are also cited as US Army H-34As, except for 1003, which was a US Navy HSS-1N.

On the port side, SA-127 carried a second missile and six 5" HVAR. Shields protected the main gear from missile exhaust. (Photo B87-1604 courtesy of SHAA)

Chapter 11

Germany

Air Force

The first of seven H-34G.Is purchased for the *Luftwaffe* (German Air Force) arrived on 25 July 1957, the last on 10 February 1958. Deliveries were initially made to *Flugzeugführerschule* (Pilot Training School) "S". Aircraft moved between that unit, the *Hubschrauberführerschule* (Helicopter Pilot School), the *Flugbereitschaft* (Executive Flight Detachment), *Hubschraubertransportgeschwader* (Helicopter Transport Squadron) 64, the *Hubschrauber-Lehr- und Versuchsstaffel* (Helicopter Training and Trials Squadron), and *1. Luftrettungs- und Verbindungs-Staffel* (First Air Rescue and Communications Squadron). These H-34s were joined by four transferred from the Army (which received two of the *Luftwaffe* Choctaws). A pair of H-34G.IIs joined the *Flugbereitschaft*, one in late July 1959 and the other in late September of the same year. Four H-34G.IIIs also came straight to the Air Force, three in late 1962 and the last in mid-1964. Two of these were transferred to the Army, from which came two others. Of these aircraft, only one H-34G.I (c/n 58-583) and one H-34G.III (c/n 58-1525) were written off before the type was retired.

As the unit names suggest, *Luftwaffe* H-34s undertook a variety of duties despite their small numbers, and these roles called for modifications. Executive Flight Detachment aircraft sported extra glazing and VIP interiors, while COMINT (communications intelligence) duties required special antennae suites. As was the case with many of the services in which it was found, the Choctaw provided capabilities that had not existed before. While it began to give way to Bell UH-1Ds in February of 1968, barely ten years after its arrival, the last were not retired for another six years.

Army

Extensive use of the H-34 by the *Heerenflieger* (Army Aviation) began with *Heeresfliegertransportstaffel* (Helicopter Transport Squadron) 823 in 1957. Deliveries of nineteen H-34G.Is ran from 18 November 1957 through 25 June 1958 (two went to the *Luftwaffe*, which transferred two to the *Heerenfliegern* in turn). Twenty-three H-34G.IIs arrived between 21 July 1959 and 22 February 1960. Forty-eight H-34G.IIIs came from Sikorsky between 4 October 1962 and 3 September 1963 (four were transferred to the Air Force, and

H-34G.I GD+234 (c/n 58-783) of the Luftwaffe's Helicopter Transport Squadron 64. (Photo 93-1064 courtesy of Smithsonian Institution)

Army H-34G.I PJ+366 (c/n 58-856) with HFlgStff(LL) 9 while at Heidelburg in 1966. The ARA-31 FM homing antennae have been added to the clamshell doors. (Photo courtesy of Norm Taylor)

H-34G.II CA+351 (c/n 58-1099) had been fitted with VIP interior and steps by the time this picture was taken on 15 August 1996. *Flugbereitschaft BmVg* was operating from Kohn Air Base. (Photo by Nahon courtesy of Norm Taylor)

Army H-34G.III 80+58 (c/n 58-1514, BuNo 150742) without high-viz markings. (Photo courtesy of Terry Love)

Late-pattern landing gear, kicksteps, and exhaust distinguished the H-34G.IIIs. The ARA-31 antennae and the rearview mirrors are common features of German H-34s. QB+482 (c/n 58-1538, BuNo 150752) served with HFlgWaS at Buckenburg in August, 1966. (Photo courtesy of Norm Taylor)

two came from there). Three H-34G.Is (c/ns 58-854, 887, and 881), three H-34G.IIs (c/ns 58-1089, 1094, and 1106), and two H-34G.IIIs (c/ns 58-1505 and 1678) were written off during the course of the type's service with the Army. During their service life, earlier models were regularly upgraded to the standards of later arrivals. This particularly included engine and avionics changes, so that by the end of their use it was impossible to distinguish between the G.I and G.II models without reference to the aircraft code.

The ninety Army Choctaws were used for transport of troops (sixteen-seat interiors were standard) and cargo, pilot and SAR training, and medevac. Replacement by the UH-1D began in 1967, and by 1972 the last H-34Gs of the transport regiments and the Army's flight school were retired in favor of the CH-53G. Plans to equip some otherwise surplus Army H-34s with 105mm rockets did not reach fruition.

MFG5 H-34G.III 80+80 (c/n 58-1567, BuNo 150764) in the aluminum-base high-viz scheme during June, 1969. (Photo by Ben Knowles courtesy of Terry Love)

One of the few H-34G.III Marine, 82+01 (c/n 58-1590, BuNo 151729) during a training exercise with MFG5 in May, 1973. (Photo by Kurt Thomsen courtesy of Terry Love)

Navy H-34G.III WE+554 (c/n 58-1605, BuNo 150811) of MFG5 at Leck AFB on 1 June 1967. (Photo 2802 courtesy of AAHS and Denis Hughes Collections)

Navy

In Germany, the *Marineflieger* (Naval Aviation) was last to obtain Choctaws, last to retire them, and the only branch to use armed Seabats. The first three H-34G.IIIs were delivered to the *Marinefliegerdienst- und Seenotgeschwader* (Marine Flight Duty and Rescue Squadron) on 22 October 1962, with nine more delivered between 27 March and 10 June of 1963. Ten were eventually transferred to *Marinefliegergeschwader* (MFG, Navy Air Squadron) 5, which also received

the thirteenth and final H-34G.III to roll off Sikorsky's production line on 14 October 1967. Nine more were received from the Army. Four H-34G.IIIs were lost in accidents (c/ns 58-1568, 1588, 1631, and 1665).

The armed Seabats were a batch of five H-34G.III Marine, delivered from 15 April 1963 to 4 August 1964, which were passed on to MFG 4 along with two H-34G.IIIs. These were equipped with AN/AQS-5 sonar and Mk.43 acoustic homing torpedoes for the ASW role.

H-34G.III 80+97 (c/n 58-1631, BuNo 150818) minus SAR markings, May, 1972. (Photo courtesy of John R. Kerr)

The green base color was unusual on Navy H-34G.IIIs, such as 81+04 (c/n 58-1671, BuNo 150802) of MFG5, seen on 12 June 1969. (Photo courtesy of CFAP)

H-34G.III QK+584 (c/n 58-1679, BuNo 150807) with HFlgTrspStff(San) 855 at Buckeburg on 12 August 1966. (Photo courtesy of Norm Taylor)

In addition to the SAR and ASW roles, a number of H-34G.IIIs assigned to MFG 4 were equipped with heavy towing gear for minesweeping. Negotiations with Sikorsky in 1971 to convert six Navy Choctaws to turbine power were not completed, and the last of the type was replaced in MFG 5 by a Sea King Mk.41 on 1 April 1975.

H-34G.I
26, equivalent to H-34As; c/ns were 58-: 532, 583, 690, 701, 721, 727, 740, 748-750, 782-783, 801-802, 827-828, 833-834, 854-857, 879, 881-883

H-34G.II
25, equivalent to H-34As with addition of ASE; c/ns were 58-: 1089 to 1112, 1135

H-34G.III
65, equivalent to H-34Cs built on HSS-1N airframes; c/ns (and BuNos originally assigned) were 58-: 1458 to 1459(150733-150734), 1493(150735), 1502-1505(150736-150739), 1512-1516(150740-150744), 1523-1527(150745-150749), 1536-1539(150750-150753), 1547-1548(150754-150755), 1553(150760), 1557(150814), 1561-1569(150761-150766, 150808-150810), 1570(150767), 1575-1578(150768-150771), 1582-1584(150772-150774), 1588-1589(150815-150816), 1594(150775), 1596(150777), 1605(150811), 1617-1618(150812-150813), 1630-1632 (150817-150819),

1658(150797), 1662-1665(150798-150801), 1671-1673(150802-150804), 1677-1679(150805-150807), 1732(152188), 1813(155290)

H-34G.III Marine
5, equivalent to SH-34Js; c/ns (and BuNos originally assigned) were 58-: 1590(151729), 1602(151730), 1619(151731), 1733(152380), 1737(152381)

Of the 121 H-34s received by Germany, fourteen were destroyed in accidents, seven were scrapped, and six went to museums. The remaining ninety-four entered civil registries around the world.

Heerenflieger
Aircraft not specifically referred to below came to the Army directly from Sikorsky. In addition, the following were transferred from the Luftwaffe:
    H-34G.I: 58-690, 58-727
    H-34G.III: 58-1526, 58-1561

Luftwaffe
H-34G.I: Delivered from Sikorsky were c/ns 58-532, 583, 690, 701, 727, 782, and 783; 801, 802, 882 and 883 were transferred from the Army.
    H-34G.II: c/ns 58-1093 and 1099 were delivered from Sikorsky.

H-34G.III Marine 82+04 (c/n 58-1733, BuNo 152380) on display in mid-1977, long after retirement from military service. (Photo courtesy of MAP via Bob Burns)

H-34G.III: c/ns 58-1524, 1526, 1561, and 1732 were delivered from Sikorsky; 58-1493 and 1525 were transferred from the Army.

Marineflieger
H-34G.III: c/ns 58-1557, 1567 to 1569, 1588 and 1589, 1605, 1617 and 1618, 1630 to 1632, and 1813 were delivered from Sikorsky; 1553, 1562, 1582, 1662 and 1663, 1665, 1671, 1673, and 1677 were transferred from the Army.
H-34G.III Marine: All were delivered from Sikorsky.

Beginning on 13 November 1967, there was a temporary application of a universal code system for German H-34s (80+nn, 81+nn, and 82+nn). This system makes identification of operating units, and therefore branches, difficult during the first half of 1968. It is possible that transfers and loans not listed above took place during that time.

Chapter 12
Haiti

The Haitian Air Force operated at least seven H-34s as transports out of Bowen Field from the late 1970s through the early 1990s. Some are believed to have come from American military surplus stocks, while others were ex-German machines sold from the American civil register after military service.

At least three (H-1, H-7, and H-10) were converted to turbine powerplants by Orlando Helicopters in Florida before their service in Latin America. H-10 had originally been refurbished by Orlando Helicopters with a VIP interior for Presi-dential service in Nicaragua. It was transferred to Haiti after the fall of the Somoza regime.

Known Haitian H-34s are:

model	c/n	s/n	notes
S-58ET	58-1526	H-7	ex-H-34G.III 150748
S-58ET	58-1536	H-1	ex-H.34G.III 150750
S-58ET	58-1583	H-10	ex-H.34G.III 150773

Known additional serials were H2 to H4, and H8. H8 might have been a former USAF HH-34J.

The press examines S-58ET H-1 (c/n 58-1536, ex-H-34G.III 150750) at Bowen Field in 1980. (Photo by George Kemp courtesy of Dan Hagedorn)

Still a "round-nose" when at Miami in October, 1978, H-8's identity is unknown. Several HH-34Js (c/ns 58-1288, 1297, 1313, 1325, and 1326) are likely candidates due to a lack of post-USAF history. (Photo by ALPS courtesy of Dan Hagedorn)

Haitian S-58ET H-7 (c/n 58-1526, ex-H-34G.III 150748) in 1992. (Photo courtesy of Jake Dangle)

S-58ET H-10 (c/n 58-1583, ex-H-34G.III 150773) with troop seats visible behind the custom glazing. It served previously as a VIP transport in Nicaragua. (Photo courtesy of Jake Dangle)

Chapter 13
Indonesia

The first H-34 operated by the *Tentara Nasional Indonesia - Angkatan Udara* (TNI-AU, Indonesian Air Force) was a VH-34D produced for sale via MDAP. This aircraft was delivered on 29 November 1961. It was serialled M251 (later changed to H-3451) and used in the VIP transport role.

An additional nine H-34s were next obtained for use as cargo and troop transports. All were ex-Marine Corps UH-34Ds. These were assigned to the Seventh Helicopter Squadron at Semplak AFB, West Java. When the Seventh transitioned to Hueys, the H-34s were used to replace the Soviet-supplied Mi-4s of the Sixth Helicopter Squadron. A later transfer moved the fleet to the Third Helicopter Squadron at Atang Senjaya AFB in Bogor, West Java.

Three more aircraft were later acquired from surplus stocks at Davis-Monthan. These were a pair of ex-Army CH-34Cs, and an ex-Navy UH-34J that had already seen service with the Vietnamese Air Force.

In 1978, California Helicopter upgraded the TNI-AU fleet to S-58Ts with the installation of Pratt & Whitney PT6T-3 TwinPac turbines. At the same time, the UH-34D interiors were modified from twelve-troop to sixteen-troop seating. There were no reported avionics upgrades at that time.

Along with SAR, medevac, and general transport duties, the S-58Ts have served as the prime mover for Indonesia's *Kopaskhasau* (Special Forces).

Known Indonesian H-34s are:

model	c/n	s/n	notes
UH-34G	58-141		ex-VNAF 138462
S-58T	58-381	H-3411	ex-CH-34C 54-3028
S-58T	58-427		ex-CH-34C 55-4469
S-58T	58-466		ex-UH-34D 143967
S-58T	58-515		ex-UH-34D 143982
S-58T	58-564	H-3406	ex-VNAF 143886
UH-34J	58-654		ex-VNAF, MASDC 143916
S-58T	58-950	H-3407	ex-VNAF 145782
S-58T	58-952		ex-UH-34D 145783
S-58T	58-1116	H-3408	ex-UH-34D 147175
S-58T	58-1173	H-3404	ex-VNAF, MASDC UH-34D 148060
S-58T	58-1224	H-3413	ex-UH-34D 148105
S-58T	58-1411	H-3451	MDAP VH-34D 150691
S-58T	58-1476	H-3414	ex-UH-34D 149374
UH-34D	58-1479		ex-VNAF, MASDC 149377
UH-34D	58-1629		ex-VNAF, MASDC 150247
S-58T	58-1685	H-3415	ex-UH-34D 150558

An unidentified TNI-AU S-58T undergoes maintenance at Bandung in 1978. (Photo courtesy of Jake Dangle)

S-58T H-3408 (ex-UH-34D 147175) with an extra cabin window, enlarged door window, 16-seat interior, and Third Helicopter Squadron insignia. If the reported c/n (58-1116) is correct, the v-leg main gear is a retrofit. (Photo courtesy of Jake Dangle)

Chapter 14
Israel

Following the 1956 Sinai War, a delegation from the Israeli Defense Force (IDF) met with French Air Force representatives to review the use of helicopters in the Algerian conflict. With this as a guide to Israeli requirements, a decision was reached to obtain S-58s and Sud Aviation Alouette IIIs rather than continue purchasing S-55s.

The Israeli Air Force (IDF/AF or IAF) began their use of the S-58, as they designated all models in their service, with the purchase of three aircraft. Shown on Sikorsky's allocation list as "S-58B, Israeli Government," these were militarized after receipt.

The first IAF S-58 (c/n 58-437), with export registration N9F, was assigned serial number 11. It is seen at Bridgeport before delivery. (Photo S22979-B courtesy of Sikorsky Aircraft Corporation)

Also with export r/n N9F, this is S-58B 12 (c/n 58-691). (Photo S23552 courtesy of Sikorsky Aircraft Corporation)

One of the "German" H-34G.IIIs displays French-pattern nose armor. (Photo courtesy of the author's collection)

Before restoration at the IAF Museum, H-34G.III 07 (c/n 58-1595, BuNo 150776) showed signs of deterioration. (Photo courtesy of the author's collection)

The original three aircraft were followed in April of 1960 by two H-34As, an S-58 originally used by Sikorsky as a demonstrator, and an ex-US civil S-58B. This complement of seven medium-lift helicopters represented a dramatic improvement over the troop and cargo capabilities of the earlier S-55s. These early model Choctaws served until the type's withdrawal from service in 1968, and all but one are known to have survived to the civil market.

Owing to Israeli satisfaction with the early-production aircraft, a deal was struck with the West German government to obtain twenty-four late-production airframes. Designated CH-

34As by Sikorsky and H-34G.IIIs by Germany, they were part of German war reparations to Israel. These have often been reported as coming from German military surplus, but in fact never reached Germany. Deliveries took place between November, 1962, and July, 1963. All were assigned to the 124th ("Rotor and Sword") Squadron.

IAF use of the S-58 was typical. The first rescue mission, in July, 1958, was the evacuation of an injured sailor from an Israeli ship; this mission involved a landing on an American carrier when the S-58 ran low on fuel. Prior to the Six Day War, the type was noted and praised for its work in both civil-

H-34G.III 21 during a take-off roll. Tan paint has been applied over the green base to provide camouflage. (Photo courtesy of the author's collection)

ian and military medevacs, the supply of isolated settlements, fire fighting, and SAR.

During the 1967 Six Day War, IAF S-58s moved troops into key positions behind Egyptian and Syrian lines. Such movements had been practiced the previous year, when a complete brigade of paratroops was moved in a relay of twenty helicopters. In addition to the troop movements, 680 casualties were evacuated and thirteen downed pilots recovered.

As with the earlier models, the "German" S-58s continued in use until 1968. Only fourteen survived military service to enter the civil market. Of the remaining ten, one serves as a display piece at the IAF Museum, and five were reportedly placed in storage upon retirement. Thus, of a total of thirty-one aircraft, only five were lost from all causes in a ten-year period. Two losses are known to have been accidental, and one due to ground fire.

There are reports of some of the stored S-58s being turned over to the United Nations for operations in Nigeria in 1969, but no identities have been determined and the rumor is denied in Israel.

first batch:

model	c/n	s/n	notes
S-58B	58-437	11	to civil S-58BT
S-58B	58-691	12	lost in accident
S-58B	58-692	13	to civil S-58BT

second batch:

model	c/n	s/n	notes
S-58	58-279	15	ex-Sikorsky N740A; to civil S-58E
S-58B	58-403		to civil S-58B
H-34A	58-1167	16	to civil S-58
H-34A	58-1186	18	to civil S-58JT

third batch (only one match to IAF serials known; BuNos were routinely assigned to aircraft built for Germany):

model	c/n	BuNo	notes
H-34G.III	58-1549	150756	
H-34G.III	58-1550	150757	
H-34G.III	58-1551	150758	to civil S-58ET
H-34G.III	58-1552	150759	to civil S-58ET
H-34G.III	58-1595	150776	s/n 07, to IAF Museum
H-34G.III	58-1606	150778	to civil S-58ET
H-34G.III	58-1607	150779	to civil S-58E
H-34G.III	58-1612	150780	to civil S-58ET
H-34G.III	58-1613	150781	to civil S-58ET
H-34G.III	58-1614	150782	to civil S-58T
H-34G.III	58-1624	150783	
H-34G.III	58-1625	150784	
H-34G.III	58-1626	150785	to civil S-58T
H-34G.III	58-1627	150786	to civil S-58E
H-34G.III	58-1636	150787	
H-34G.III	58-1637	150788	to civil S-58ET
H-34G.III	58-1638	150789	
H-34G.III	58-1639	150790	to civil S-58ET
H-34G.III	58-1646	150791	to civil S-58ET
H-34G.III	58-1647	150792	
H-34G.III	58-1648	150793	to civil S-58E
H-34G.III	58-1649	150794	to civil S-58T
H-34G.III	58-1656	150795	
H-34G.III	58-1657	150796	to civil S-58ET

C/ns 58-1658 and 1659 (BuNos 150797 and 150257) have been reported as used by the IAF; they actually saw German and US Marine Corps service, respectively.

The S-58 at the IAF Museum is often reported as ex-AdA 58-340, which AdA records show was destroyed in French service. Further, significant features such as the landing gear and kicksteps were introduced too late in the production series for 58-340 to be a reasonable identity. The identity given above is taken from data plates on the aircraft, and matches the salient features.

In the foreground during a training exercise, H-34G.III 28 is fitted with the mount for a long-range fuel tank. (Photo courtesy of State of Israel GPO via Tsahi Ben-Ami)

The Israel censor has further obscured the already vague serial of this H-34G.III. (Photo courtesy of State of Israel GPO via Tsahi Ben Ami)

Chapter 15
Italy

Italian Navy use of the H-34 spanned two decades. Ordered in 1958 as ASW aircraft, with AQS-4 sonar and Mk.43 torpedoes, the first two HSS-1s were turned over to *1° Gruppo Elicotteri* (First Helicopter Group) at Catania Fontanarossa in April, 1959. During 1962, the Group relocated to the new Naval Air Station at Maristaeli Catania. By mid-1963, a total of five HSS-1s and three HSS-1Ns had been received. They were coded 4-01 through 4-08. Following

American practice, these were redesignated SH-34Gs and SH-34Js, respectively.

On 31 October 1964, a tornado collapsed the Group's hangar and destroyed five aircraft (4-01, 4-02, 4-06, 4-07, and 4-08). Two of the survivors were at sea aboard ITS DORIA, and one was at an overhaul facility.

During 1966 and 1967, six SH-34Js were delivered. Codes assigned were 4-09 through 4-14. A crash near Au-

SH-34Gs 4-07 (c/n 58-710, BuNo 143940) and 4-06 (c/n 58-599, BuNo 143899) in close formation with SH-34J 4-03 (c/n 58-1666, BuNo 150821) over the Italian countryside. (Photo courtesy of Italian Navy)

gusta on 19 May 1967 claimed 4-11. 4-12 went down in the same area on 5 July 1968.

In 1969, the seven aircraft still in service were transferred to *5° Gruppo Elicotteri* at Maristaeli Luni. By this time, they were no longer active in the ASW role. Instead, they were engaged in SAR and troop transport missions, including the insertion of commando patrols. 4-09 crashed near Capo San Terencio on 28 November 1969. June, 1972, saw the demise of 4-05 near San Marinello. 4-13 was withdrawn from use in 1973 following a ground accident at Maristaeli Luni.

Between 1974 and 1979, the last four Italian Navy H-34s were returned to *1° Gruppo Elicotteri*, which had redeployed to Maristaeli Luni. 4-03 crashed on 18 October 1974 near Campalecina.

In June of 1979, 4-04, 4-10, and 4-14 were retired. The first two went into storage, while 4-14 was placed on display at Maristaeli Luni.

This delivery shot of HSS-1N 4-04 (c/n 1394, BuNo 149082) shows the Seabat's sonar well. The oil cooler area behind the clamshell doors has been sealed against salt water during transit. (Photo courtesy of Italian Navy)

model	c/n	s/n	code	notes
HSS-1	58-599	MM143899	4-06	destroyed in tornado
HSS-1	58-710	MM143940	4-07	destroyed in tornado
HSS-1	58-745	MM143949	4-08	destroyed in tornado
HSS-1	58-990	MM80163	21, 4-01	destroyed in tornado
HSS-1	58-991	MM80164	22, 4-02	destroyed in tornado
HSS-1N	58-1394	MM149082	26, 4-04	to storage
HSS-1N	58-1395	MM149083	27, 4-05	crashed
SH-34J	58-1666	MM80237	23, 4-03	to MM150821; crashed
SH-34J	58-1770	MM153617	4-09	crashed
SH-34J	58-1771	MM153618	4-10	to storage
SH-34J	58-1772	MM153619	4-11	crashed
SH-34J	58-1773	MM153620	4-12	crashed
SH-34J	58-1774	MM153621	4-13	ground accident
SH-34J	58-1775	MM153622	4-14	base display

All aircraft in Italian military service have the serial prefix MM (*Matricola Militare*, figuratively Military Serial). Except for 4-01, 4-02, and the original 4-03 serial, the above numbers represent US Navy BuNo assignments, common for MAP and MDAP aircraft.

This list is based on material received from the Italian Navy, and is consistent with reports from Italian researchers. The Sikorsky allocation list, however, shows an additional five aircraft as built for Italy for which there are no Italian records, serials, or photographs:

model	c/n	BuNo
HSS-1N	58-1404	149084
HSS-1N	58-1405	149085
HSS-1N	58-1417	149086
HSS-1N	58-1418	149087
SH-34J	58-1667	150822

...ound are HSS-1Ns 4-13 (c/n 58-1774, BuNo 153621) and 4-05 (c/n

Chapter 16
Japan

Eight HSS-1s were taken on charge by the Japanese Maritime Self-Defense Force (JMSDF) between 1 April 1958 and 29 March 1961. Assigned JMSDF serials 8551 through 8558, they first served with Helicopter Anti-submarine Squadron 101 at Naval Air Station Tateyama. 8554 was stricken in late 1960, 8552 and 8558 in 1963, and 8555 in 1968; the remaining four stayed on inventory into the early 1970s.

HAS-101's Seabats were supplemented by nine HSS-1Ns. Produced as components by Sikorsky, and assembled by Mitsubishi, these became serials 8561 through 8569. Deliveries occurred between 31 August 1961 and 26 January 1963. Only one, 8567, was lost during the type's service in Japan, this to a crash on 7 April 1969.

In 1963, the HSS-1s were transferred to the Ominato Air Squadron, while the HSS-1Ns remained with HAS-101. From 1966 until retirement in 1971, the early models served with the Omura Air Squadron. During this period, the late models were with the Komatsujima Air Squadron. They moved to the Omura Air Squadron in 1971, and were retired from there in 1973.

Except for 8569, which went on display at Tateyama, all of the JMSDF Seabats were scrapped after retirement.

model	c/n	s/n
HSS-1	58-808	8551
HSS-1	58-835	8552
HSS-1	58-1143	8553
HSS-1	58-1144	8554
HSS-1	58-1237	8555
HSS-1	58-1238	8556
HSS-1	58-1248	8557
HSS-1	58-1249	8558
HSS-1N	58-1333	8561
HSS-1N	58-1349	8562
HSS-1N	58-1358	8563
HSS-1N	58-1359	8564
HSS-1N	58-1368	8565
HSS-1N	58-1379	8566
HSS-1N	58-1528	8567
HSS-1N	58-1529	8568
HSS-1N	58-1543	8569

Two HUS-1s were produced for the Japanese civil register in August, 1958. These were c/ns 58-945 and 946, which were registered JA7201 and JA7202. While military models, there are no records for these of military service. JA7201 is on display in the National Science Museum.

8551, the first JMSDF HSS-1 (c/n 58-808) in an early all-gray scheme. (Photo S24390-A courtesy of Sikorsky Aircraft Corporation)

HSS-1N 8564 (c/n 58-1359) hovers near the tower at Omura Air Base. JMSDF H-34s retained the low-stack exhaust of the R1820-84B engine. (Photo courtesy of the author's collection)

8568 (c/n 58-1529) with weapons racks, dunking sonar, and Doppler radar heads. (Photo courtesy of the author's collection)

Although a HUS-1 Seahorse, JA7201 (c/n 58-945) never saw military service. (Photo courtesy of Nick Williams)

Chapter 17

Katanga

Two S-58Cs were leased by Sabena, which operated them as short-haul passenger carriers, to *l'Aviation Katangaise* (Avikat, the Katangan Air Force). These were assigned serials KAT43 and 44. The serials were apparently never painted on the aircraft, which retained their Sabena livery and registrations. Obtained in February, 1961, they served briefly as transports during raids against the Congo but were used mainly as medevacs. It is unclear how long they remained with Avikat, but both had returned to Sabena and been transferred to the Belgian Air Force by October, 1963.

model	c/n	s/n	r/n
S-58C	58-324	KAT43	OO-SHG
S-58C	58-395	KAT44	OO-SHM

S-58C OO-SHG (c/n 58-324), with one of its S-55 predecessors, before serving in Katanga as KAT43 and with the Belgian Air Force as OT-ZKI/B9. (Photo courtesy of Rudy Binnemans)

OO-SHM (c/n 58-395) flew for a short time as KAT44, without changing markings, and became LuM OT-ZKM/B13. (Photo courtesy of Rudy Binnemans)

Chapter 18
Laos

The Royal Laotian Air Force received the first of its UH-34Ds in May, 1963, following a period of training conducted in Thailand by US Marines. Seahorses came to the RLAF after service in Laos with Air America, from the Vietnamese Air Force, and from the *Commission Internacional de Control* (CIC, International Control Commission) stationed in Laos under the Geneva Accords. Maintenance was provided by Air America.

While it is certain that the Laotian H-34s served in the fighting against the Pathet Lao and North Vietnamese Army as cargo and troop transports, as well as medevacs, nothing certain is known of the units they equipped or their dates of service. There have been reports of RLAF H-34s in service into the late 1970s. Eight of the known aircraft are presumed to have been lost due to accidents or combat, but the date of only one loss is available. Even the precise number operated is unclear.

Twenty-one of the RLAF (and ex-VNAF) UH-34Ds were transferred to Thailand in the late 1960s. One went to Air America, and two to the Philippine Air Force.

Ex-AAM UH-34D 145769 (c/n 58-908) was one of the aircraft rebuilt by AAM at Udorn for RLAF service, as indicated by the late-pattern kicksteps mixed with bent-leg main gear. In the background is ex-VNAF UH-34D 144637 (c/n 58-537). (Photo by James Fore courtesy of Larry Jacks)

Known RLAF H-34s are:

model	c/n	s/n	notes
UH-34D	58-537	144637	ex-VNAF
UH-34D	58-908	145769	ex-AAM
UH-34D	58-964	145786	ex-Air America
UH-34D	58-1388	148803	ex-CIC; lost 25 October 1970
UH-34D	58-1390	148805	ex-CIC
UH-34D	58-1391	148806	ex-CIC
UH-34D	58-1392	148807	ex-CIC
UH-34D	58-1415	149320	ex-Air America
UH-34D	58-1615	63-8248	ex-VNAF; to Thailand
UH-34D	58-1640	63-8249	ex-VNAF; to Thailand
UH-34D	58-1641	63-8250	ex-VNAF; to Thailand
UH-34D	58-1642	63-8251	ex-VNAF; to Thailand
UH-34D	58-1653	63-8252	ex-VNAF; to Thailand
UH-34D	58-1654	63-8253	ex-VNAF; to Thailand
UH-34D	58-1655	63-8254	ex-VNAF; to Thailand
UH-34D	58-1686	63-8255	ex-SHAA; to Thailand
UH-34D	58-1687	63-8256	ex-VNAF; to Thailand
UH-34D	58-1701	63-8257	ex-VNAF; to Thailand
UH-34D	58-1702	63-8258	ex-VNAF; to Thailand
UH-34D	58-1714	63-8259	ex-VNAF; to Thailand
UH-34D	58-1726	63-13006	ex-VNAF; to Thailand
UH-34D	58-1727	63-13007	ex-VNAF; to Thailand
UH-34D	58-1728	63-13008	ex-VNAF; to Thailand
UH-34D	58-1729	63-13009	ex-VNAF; to Thailand
UH-34D	58-1730	63-13010	ex-VNAF; to Thailand
UH-34D	58-1731	63-13011	ex-VNAF; to Thailand
UH-34D	58-1734	63-13012	ex-VNAF; to Thailand
UH-34D	58-1735	63-13013	ex-VNAF; to Thailand
UH-34D	58-1736	63-13014	ex-VNAF; to Thailand
UH-34D	58-1744	63-13139	ex-VNAF; to Philippines
UH-34D	58-1747	63-13140	ex-VNAF; to Philippines
UH-34D	58-1801	154891	to Air America

154891 (c/n 58-1801), built in 1967, was one of the last UH-34Ds produced. It went on to AAM. The other aircraft is ex-AAM UH-34D 145786 (c/n 58-964).(Photo courtesy of Air America Archives, History of Aviation Collection, University of Texas at Dallas)

The "to and from" of RLAF H-34s is a matter of speculation. There are no official records available, and most photos of RLAF Seahorses are undated. Some sources suggest that the aircraft in the 63-82nn and 63-13nnn ranges went to Laos first and then to Vietnam. It seems equally likely that they took the opposite route, as Vietnam replaced its H-34s with Hueys during the same period that the war in Laos was expanding. This would fit with the recollections of Air America crews who served in Laos. The production dates for these Seahorses are also consistent with the activation of new VNAF helicopter squadrons during the first half of the 1960s.

RLAF HSS-1 148805 (c/n 58-1390) first served with AAM as H-Y and H-48, and with the CIC. It did not, however, survive RLAF service. (Photo B87-2066 courtesy of SHAA)

Chapter 19
Netherlands

Twelve HSS-1Ns were purchased to equip 8 Squadron of the *Koninklijke Marine* (Royal Netherlands Navy). Five were taken on strength on 5 January 1960. The first was coded 8-5, after which the batch was coded H-5 through H-9. Later, the codes became 141 through 149. Another three followed on 4 September 1961 (138-140), two on 30 March 1962 (136-137) and the final pair on 26 June 1963 (134-135). Royal Netherlands Navy Seabats served aboard

HSS-1N 142 (c/n 58-1150, BuNo 147632) with 7 Squadron at Valkenburg prior to September, 1962. The sonar cabinet is visible in the cabin. (Photo courtesy of the author's collection)

SH-34J 143 (c/n 58-1153, BuNo 147633) of the Karel Doorman-embarked 8 Squadron on 5 July 1967. It became a UH-34J with 7 Squadron before the end of the year. (Photo courtesy of CFAP)

HSS-1N 138 (c/n 58-1373, BuNo 149131) shortly after its 4 September 1961 delivery. (Photo 6-28 courtesy of Royal Netherlands Navy via Dana Bell)

HMS KAREL DOORMAN while with 8 Squadron; the only other countries to mount regular carrier operations with their H-34s were the United States, France, and Brazil.

In 1967, the *Marine Luchtvaart Dienst* (MLD, Naval Aviation Service) stripped the aircraft of ASW gear. Redesignated UH-34Js, they were transferred to 7 Squadron at Valkenburg Air Base. By this time, five had been written off. 139 ditched at Moray Firth, near Scotland, on 25 October 1961. 140 and 145 were ditched near Gibraltar on 31 January 1963. 136 was lost near Noordwijk, in the Netherlands, on 17 August 1965. Finally, in terms of losses, 141 went into the ocean off Scotland on 22 September 1965. Although not stricken, 142 was heavily damaged while being moved by elevator to the KAREL DOORMAN's flight deck on 14 September 1963; re-

pairs were undertaken by Henschel. Meanwhile, 143 and 144 collided on 28 December 1965, and were rebuilt by the Naval Aircraft Repair Works and Henschel, respectively.

During the course of service at Valkenberg, a further four were stricken from inventory. 143 was scrapped on 4 December 1969 after a ground resonance accident, and 135 followed on 24 June 1970. On that same day, 138 and 144 were declared surplus. Both went to MAAG Netherlands for sale, and found their way to American civil use.

The remaining three were retired on 28 June 1972. Like 138 and 144, 137 and 142 were sold to US firms by MAAG Netherlands. 134 went to the Anthony Fokker School at The Hague as an instructional airframe, and then to the Military Aviation Museum.

8 Squadron SH-34J 140 (c/n 58-1375, BuNo 149133) during a training exercise from Karel Doorman in September, 1962. It was lost near Gibraltar on 31 January 1963. (Photo 6-30 courtesy of Royal Netherlands Navy via Dana Bell)

model	c/n	code	BuNo	notes
HSS-1N	58-1145	141	147631	was 8-5, H-5; lost
HSS-1N/UH-34J	58-1150	142	147632	was H-6; sold
HSS-1N/UH-34J	58-1153	143	147633	was H-7; scrapped
HSS-1N/UH-34J	58-1158	144	147634	was H-8; sold
HSS-1N	58-1160	145	147635	was H-9; lost
HSS-1N/UH-34J	58-1373	138	149131	sold
HSS-1N	58-1374	139	149132	lost
HSS-1N	58-1375	140	149133	lost
HSS-1N	58-1430	136	149841	lost
HSS-1N/UH-34J	58-1442	137	149842	sold
HSS-1N/UH-34J	58-1597	134	150729	to Fokker School
HSS-1N/UH-34J	58-1611	135	149840	scrapped

Chapter 20
Nicaragua

Starting in mid-1975, the *Fuerza Aérea Nicaragua* (FAN, Nicaraguan Air Force) operated a dozen surplus H-34s of various types. Deliveries of aircraft taken from storage at Davis-Monthan continued sporadically until mid-1977. Serials assigned were FAN517 through FAN527 (FAN519 was used on two different aircraft). The FAN H-34s were used as troop and cargo transports during the country's civil wars, with at least one being reported fitted with light bomb racks under the fuselage. At least three were converted to S-58Ts. Two (FAN517 and the first FAN 519) were destroyed in crashes. Those that survived until the Sandinista takeover were scrapped in favor of Russian equipment by the *Fuerza Aérea Sandinista*, which briefly operated FAN518 as FAS-51.

A thirteenth example operated as President Samoza's executive transport, and carried a civil registration. It found its way to Haitian military service by way of Florida after Samoza fled the country in it.

Known Nicaraguan H-34s are:

model	c/n	s/n	notes
CH-34A	58-365	FAN526	ex-54-3019
CH-34A	58-480	FAN527	ex-55-4488
CH-34A/S-58T	58-687	FAN518	ex-56-4312; to FAS-51
CH-34A	58-702	FAN522	ex-56-4313
CH-34A	58-719	FAN517	ex-56-4321; crashed 9/9/76
CH-34A/S-58T	58-843	FAN520	ex-57-1699
CH-34A/S-58T	58-844	FAN519(1)	ex-57-1700; crashed 11/17/78
HSS-1N	58-886	FAN519(2)	ex-145673
HSS-1N	58-987	FAN521	ex-145692
S-58ET	58-1583	YN-PLN	ex-H-34G.III 150773

While original identities are not known for FAN523 through FAN525, photographs indicate they were early production H-34As with prior US Army service.

CH-34C FAN526 (c/n 58-365, ex-USA 54-3019) at Tucson, AZ, on 3 May 1977. (Photo by Ben Knowles courtesy of Dan Hagedorn)

FAN518 (c/n 58-687, ex-USA 56-4312), a CH-34C converted to an S-58T, at Davis-Monthan in July of 1975. It later served the Sandinistas as FAS-51. (Photo by Ben Knowles courtesy of Dan Hagedorn)

The first of two aircraft serialed FAN519 (c/n 58-844, ex-USA CH-34C 57-1700) at Tucson on 26 October 1976. (Photo by Ben Knowles courtesy of Dan Hagedorn)

FAN521, an ex-USN HSS-1N (c/n 58-987, BuNo 145692) at Tucson in September, 1975. (Photo by Ben Knowles courtesy of Dan Hagedorn)

Chapter 21
Philippines

The Philippine Air Force (PAF) reports acquiring six UH-34Ds during the 1960s via MAP, although serials are only known for four. Two appear to have been ex-Laotian aircraft. Referred to as "Silver Grasshoppers" because of their final, overall aluminum color scheme, PAF H-34s were used by the 505th Air Rescue Squadron until October, 1974. Along with SAR duties, the PAF H-34s carried out aerial photogrammetry, mapping, and intelligence gathering. One was tested with air-to-ground rockets as well as bomb racks.

Although there is no information on losses, it has been reported that three were scrapped. Only one survives, and this is on display at Villamor (formerly Nichols) Air Base. Although sales were handled through the US Air Force, the known serials are a mix of USAF and Navy numbers.

Known PAF H-34s are:

model	c/n	s/n	notes
UH-34D	58-1744	63-13139	ex-Laotian
UH-34D	58-1747	63-13140	ex-Laotian
UH-34D	58-1769	153131	to PAF Museum
UH-34D	58-1776	153132	

Nose armor and OD camouflage tell of 139's earlier service with the RLAF as UH-34D 63-13139 (c/n 58-1744). (Photo courtesy of the author's collection)

UH-34D 153132 (c/n 58-1776) armed for counter-insurgency strikes with four bazooka tubes and two bombs. (Photo courtesy of the author's collection)

At the PAF Museum, UH-34D 153131 (c/n 58-1769) keeps company with a Texan and a Trojan. (Photo by E. de Kruyff courtesy of the author's collection)

Chapter 22
Soviet Union

When Soviet Premier Nikita Khruschev visited the United States in 1959, he had the opportunity to fly in one of the Marine Corps HUS-1Zs used as part of President Eisenhower's fleet. His favorable impressions of the aircraft led to the sale of two S-58s to the Soviet Union. These were delivered during the first week of December, 1960. One (c/n 58-1241) was equipped to HUS-1Z standards, and the other might have been in the cargo/troop-carrying configuration.

Only guesswork is available regarding their history following delivery. It is generally felt that one, and possibly both, went to the Mil design bureau. There, they were disassembled for study and the parts subsequently placed in storage. Serials CCCP-27491 and 27492 are known to have been assigned, but there are no other indications of civil or military ownership and use.

HUS-1Z 27492 (c/n 58-1242) during trials at Bridgeport. (Photo 30907-B courtesy of Sikorsky Aircraft Corporation)

model	c/n	s/n	notes
HUS-1Z	58-1241	CCCP-27491	carried US r/n N74162
HUS-1Z	58-1242	CCCP-27492	

In VIP configuration and carrying export registration N74162, HUS-1Z 27491 (c/n 58-1241) sits at Sikorsky's Bridgeport plant. (Photo 30889-A courtesy of Sikorsky Aircraft Corporation)

Chapter 23
Thailand

As part of an overall military buildup due to the growing war in Vietnam, Thailand had obtained eleven H-34s by late 1962. The Royal Thai Air Force (RTAF) assigned them the designation H.4. After a lengthy period of training, and many teething problems, the type became operational in 1965. The number in service rose to forty by the end of 1970, and peaked at sixty-four. These included a large portion of the Royal Laotian Air Force's inventory of H-34s. The H.4 equipped 31 and 33 Squadrons at Korat, where they played a major role in counter-insurgency operations. Seven also served with the Flying Training School as part of 63 Squadron at Don Muang.

An eighteen-aircraft mix of former US Army CH-34Cs, USMC UH-34Ds, and RLAF UH-34Ds was converted to S-58Ts by Thai-Am Incorporated from 1978 to 1980. Additionally, the Royal Thai Agricultural Air Division (Thai Ag) either transferred or loaned two S-58Ts to the RTAF. This significantly lengthened the type's Thai service life, and fourteen of the twenty S-58Ts were still in use in 1997. These flew with 201 ("Spider") Squadron at Kokkathium Air Base near Lop Buri. With turbine engines, the type was designated H.4K by the RTAF.

One of the former Thai Ag S-58Ts, number 912, was involved in an unusual trade for another airframe. The crew chief of 912 was killed in action. Believing the aircraft to be haunted by his spirit, the Thais refused to fly it. An exchange was arranged for an unidentified, earlier-model S-58T that carried the color scheme, identity plates, and serial of its predecessor. The original 912 was last reported in Florida in non-flying condition.

The RTAF regularly conducted Search and Rescue missions over land areas. The main SAR helicopter was the H.4K, and it proved well suited for such work. In 1996, the RTAF was ordered to prepare for over-water SAR missions. The performance limitations of the S-58T, and the necessary modifications to the standard operating procedures for over-water operations, were being reviewed in a study scheduled for release in mid-1997.

Known RTAF H-34s are:

model	c/n	s/n	notes
CH-34C	58-236	20153	ex-USA 54-933, to S-58T
CH-34C	58-237	20151	ex-USA 54-934, to S-58T
CH-34C	58-349	20161	ex-USA 54-3010, to S-58T
UH-34D	58-572		ex-CIC 6
CH-34C	58-1019	20157	ex-USA 57-1763, to S-58T
S-58ET	58-1117		ex-AdA 58-1117, ex-Thai Ag 911
UH-34D	58-1292	20148	ex-VNAF 148756, to S-58T
UH-34D	58-1318		ex-USMC 148767, to S-58T
UH-34D	58-1440	20172	ex-VNAF 149341, to S-58T
UH-34D	58-1480	20171	ex-VNAF 149378, to S-58T
UH-34D	58-1558	20147	ex-USMC 150218, to S-58T
S-58T	58-1564		ex-H34G.III 150764, ex-Thai Ag 912
UH-34D	58-1615		ex-RLAF 63-8248
UH-34D	58-1640	20126	ex-RLAF 63-8249, to S-58T
UH-34D	58-1641		ex-RLAF 63-8250

Unidentified RTAF round-noses. The front aircraft is an ex-VNAF Choctaw. The aircraft in the back appears to have an ex-USA CH-34C's fuselage and ex-USMC or ex-VNAF H-34's tail pylon. Such mix-and-match airframes were increasingly common as the H-34s aged. (Photo courtesy of Dana Bell)

S-58T 20133 was a UH-34D (c/n 58-1655, s/n 63-8254) when received from the RLAF. (Photo courtesy of the author's collection)

UH-34D	58-1642	20132	ex-RLAF 63-8251, to S-58T
UH-34D	58-1653		ex-RLAF 63-8252
UH-34D	58-1654		ex-RLAF 63-8253
UH-34D	58-1655	20133	ex-RLAF 63-8254, to S-58T
UH-34D	58-1686	20154	ex-RLAF 63-8255, to S-58T
UH-34D	58-1687		ex-RLAF 63-8256
UH-34D	58-1701	20135	ex-RLAF 63-8257, to S-58T
UH-34D	58-1702		ex-RLAF 63-8258
UH-34D	58-1714		ex-RLAF 63-8259
UH-34D	58-1726		ex-RLAF 63-13006, to AAM

UH-34D	58-1727		ex-RLAF 63-13007, to AAM
UH-34D	58-1728		ex-RLAF 63-13008, to AAM
UH-34D	58-1729	20143	ex-RLAF 63-13009, to S-58T
UH-34D	58-1730		ex-RLAF 63-13010
UH-34D	58-1731	20141	ex-RLAF 63-13011, to S-58T
UH-34D	58-1734	20142	ex-RLAF 63-13012, to S-58T
UH-34D	58-1735		ex-RLAF 63-13013
UH-34D	58-1736		ex-RLAF 63-13014
UH-34D	58-1767	20121	ex-USMC 153129, to S-58T
UH-34D	58-1788	20173	ex-VNAF 153123, to S-58T

An RTAF T-nose in the later disruptive camouflage scheme during crew training at Lop Buri in 1984. (Photo courtesy of Jake Dangle)

The original identity of the replacement Thai Ag 912, seen in August of 1996, is unknown. The cabin glazing suggests a prior civil history, while the mix of main gear and kickstep styles puts it in the c/n range 58-1165 to 58-1393. (Photo courtesy of Steven Darke)

Chapter 24
United States

Navy

In the Navy's search for an ASW helicopter with greater range and payload than the H-19, a contract for four Seabat prototypes was issued in late June, 1952. These were designated XHSS-1s, and assigned construction numbers 58-1x through 4x. Deliveries occurred between mid-December, 1953, and late May, 1954.

The HSS-1 design was pitted in competitions against Bell's HSL-1. The latter was originally judged superior and orders for production aircraft were placed. However, it was quickly apparent that Bell's tandem-rotor design was too large for easy operation from carriers, and too noisy for the sonar operator to do his job while searching for submarines. As a

Although the 10th production H-34, 137854 was the first HSS-1 delivered and spent its life as a testbed, thus earning the YSH-34G designation after September of 1962. The color scheme is overall fluorescent red-orange. (Photo courtesy of MSgt David W. Menard, USAF, Ret.)

Reserve HSS-1 140127 (c/n 171) on static display during an open house. (Photo courtesy of MSgt David W. Menard, USAF, Ret.)

UH-34G 141586 (c/n 58-232) with HT-8 at Ellyson Field, FL, in standard trainer scheme. (Photo KN-23156 courtesy of National Archives via Rob Mignard)

Stripped of ASW gear, UH-34G 141592 (c/n 246) served at Pt. Mugu, CA, when this photo was taken on 15 May 1965. (Photo by D. Kasulka courtesy of Norm Taylor)

result, orders were placed for production HSS-1s even before the first XHSS-1's maiden flight on 8 March 1954.

Deliveries of HSS-1s began in December of 1954, and examples entered Fleet service with Helicopter Anti-submarine Squadron Three (HS-3) in August of 1955. Minesweeping trials were already being conducted with the Naval Air Mine Defense Development Unit at Panama City, Florida. The last of 187 of the early model Seabat was delivered on 5 September 1958. The first all-weather, night-capable HSS-1N had been delivered to HS-1 at Key West, Florida, a week earlier. Under Aircraft Service Change Eighty-four, 105 HSS-1s were upgraded to HSS-1N standards.

Navy orders resulting in new production of 119 of the advanced Seabat were completed with the delivery of BuNo 149842 in May, 1962. By this time, the Sikorsky HSS-2 Sea King was coming into service and, for the Navy at least, the SH-34's front-line days were drawing to a close. The last SH-

"King Pin II," VX-6's distance-record setting HUS-1L (c/n 58-545, BuNo 144657), before departure for Antarctica. (Photo courtesy of Ken Snyder)

Field green, fluorescent red-orange, and white UH-34J 143915 (c/n 58-650) with VIP interior at NAF Naples, Italy, during July of 1967. (Photo by S. Peltz courtesy of Norm Taylor)

UH-34J 143920 (c/n 58-658) at NAS Brunswick. This Seabat was one of those upgraded from HSS-1 to HSS-1N standards while in ASW configuration. (Photo Col. 1603 courtesy of Ronald W. Harrison)

HSS-1 143939 (c/n 58-709) with HS-4. The dunking sonar head is being streamed beneath the aircraft. (Photo courtesy of the author's collection)

Colorful UH-34E 145716 (c/n 58-807) with VC-5 at Naha Air Base on 15 January 1968. The tail float guard is still mounted. (Photo by Steve Miller courtesy of Terry Love)

Worn UH-34J 145682 (c/n 934) assigned to NAS Sigonella. (Photo 3789 courtesy of J.M. Friell)

Reserve HSS-1N 145681 (c/n 929) at NAS New York. (Photo by Ronald Montgomery courtesy of Kevin W. Pace)

HSS-1N 145711 (c/n 58-1064) of HS-6 while embarked aboard USS Kearsarge (CVS-33) for her 1959-1960 cruise. (Photo courtesy of Rick Albright)

SH-34J 148003 (c/n 58-1255), from NAS Pt. Mugu, at Van Nuys Airport. The original "NAVY" was crudely painted over before the joint-service Reserve markings were applied. (Photo courtesy of Nick Williams)

At yet another open house display, SH-34J 148014 (c/n 58-1272) from the Reserves at NAS Jacksonville. (Photo courtesy of the author's collection)

34J left Regular Navy service with its departure from HS-2 at NAAS Ream Field, California, in 1965. Reserve and training units continued to operate SH-34Gs and SH-34Js (most often stripped of ASW gear and redesignated UH-34G and UH-34J) in company with Marine UH-34Ds into the 1970s. "Retired" ASW H-34s also were passed along as utility aircraft to a number of friendly nations.

Navy Seabats, working as part of hunter-killer teams with destroyers and fixed-wing assets, were never called upon to sink enemy submarines, which was their original purpose. They were, however, on the scene of several military crises during their career. HS-4 (first to operate the HSS-1N in the Pacific) flew from USS PRINCETON (CVS-37) off Formosa in late 1958 and early 1959. In 1961, this time from YORKTOWN (CVS-10), HS-4 flew insertion and support missions into the jungles of Southeast Asia. HSS-1s of HS-9, embarked aboard USS ESSEX (CVS-9), were part of naval forces on the periphery of the 1961 Bay of Pigs invasion. HS-5, HS-7, HS-9, and HS-11 took part in the American block-

ade of Cuba in late 1962, operating from, respectively, the ASW carriers USS LAKE CHAMPLAIN (CVS-39), USS RANDOLPH (CVS-15), ESSEX, and USS WASP (CVS-18).

Under less bellicose conditions, Seabats flew countless military and civilian search, rescue, and relief missions around the world. Within months of reaching Fleet service, HSS-1S were ferried to Tampico, Mexico. There they delivered 197 tons of food, landed medical teams in areas cut off by floodwaters, and rescued an amazing total of 9262 civilians from flooded areas. The Connecticut floods of the same year saw a single HSS-1, piloted by Lt. G. Bello, rescue 249 people. In January of 1963, an SH-34G from Rota, Spain, operated in support of flood victims in Morocco. In a single, last-chance flight, the fourteen-passenger helicopter lifted its crew of three plus forty civilians to safety. Seabats from East Coast ASW squadrons supported NASA during spaceflight recovery operations, while scientific projects and expeditions as far from the world of war as the Galapagos Islands were aided by H-34s.

NAS Jacksonville SH-34Js 148022 (c/n 58-1285) and 147988 (c/n 58-1220). (Photo courtesy of the author's collection)

Reserve SH-34J 148948 (c/n 58-1338) from NAS Glenview visits Chanute AFB on 17 June 1965. (Photo by Vince Reynolds courtesy of John R. Kerr)

An unknown, but certainly small, number of SH-34Js was re-equipped with plush VIP interiors. These VH-34Js did not serve with the Presidential fleet, as did Army and Marine H-34s, but were used as personal transports at the higher echelons of military command.

Navy requirements also extended to the HUS-1 series, originally ordered by the Marine Corps. Thirty-three HUS-1s and UH-34Ds were delivered directly from Sikorsky between November, 1961, and May, 1967. These were joined by other Seahorses transferred from the Marines. They served as station hacks and ships' aircraft, developed the vertical replenishment practices that are now a normal part of Fleet life, were active in development and utility squadrons, and equipped Regular Navy training squadrons and Reserve units alongside their ASW counterparts. The hacks' job was not always a routine one. NAS Roosevelt Road's station UH-34D operated from the grounds of the American Embassy in the

Dominican Republic during the 1965 intervention. Armed and armored, the Seahorses assigned to LPHs off the coast of Vietnam flew regular missions in-country. In the final accounting, twenty of the forty-eight Navy helicopters lost during the Vietnam War were H-34s.

Additional Seahorses in the Navy inventory were the HUS-1A/UH-34E and the HUS-1L/LH-34D. The former version was amphibious, but three were used by VX-6 in Antarctica without their flotation bags. These paved the way for the seven factory-winterized HUS-1Ls, which served with the Antarctic development squadron until the mid-1960s. One of the HUS-1Ls delivered for Antarctic service was mistakenly equipped with a VIP interior, which was stripped before entering squadron service. With insufficient range to reach McMurdo Sound on their own, the Operation Deep Freeze H-34s were delivered in the belly of Air Force C-124 Globemasters. Of the ten specialized Navy Seahorses, a HUS-1A (BuNo 144657) and

This unidentified VH-34J, assigned to CINCLANT, has an unusually large panoramic window in the cabin door. No floats or float guards are fitted. The VIP placard holder below the cockpit is empty. (Photo 428-KN-11029) courtesy of National Archives)

Task Group Bravo/HS-11 HSS-1N 148961 (c/n 1364), with fresh day-glo and eye-catching tail stripes, is rigged for minimum-space stowage. (Photo 1493 courtesy of J.M. Friell)

NAF Monterey, CA, station hack UH-34D 149375 (c/n 58-1477). (Photo U15897 courtesy of MAP)

Beautiful UH-34Ds 149376 (c/n 58-1478) and 150208 (c/n 58-1535) from VC-1. (Photo courtesy of Jim Mesko)

a pair of HUS-1Ls (148121 and 148122) went on to the US civil market, and a single LH-34D (145717) has become a museum piece.

Direct deliveries to the Navy, by type and BuNo, were:
XHSS-1 143667-143670
HSS-1 137849-137858, 138460-138493, 139017-139029, 140121-140139, 141571-141602, 143864-143898,143900-143939, 143941-143948, 143950-143960, 145660-145670, 145672, 145678
HSS-1N 145674-145677, 145679-145712, 147984-148802, 148804, 148808-148032, 148934-148963, 149084-149087, 149841-149842
HUS-1 148822, 149318
UH-34D 152686, 153116-153120, 153124-153130, 153133, 153556-153557, 153697-153704, 154889-154894
HUS-1A 144655, 144657-144658

HUS-1L 145717, 145719, 148112, 148119, 148121-148122
LH-34D 150220

Direct deliveries to the Navy, by type, construction number (all prefixed by "58-"), and BuNo (in parenthesis), were:
XHSS-1 1x-4x(134667-134670)
HSS-1 5-14(137849-137858), 16-20(138474-138478), 24-27(138479-138482), 34-38(138483-138487), 46-49(138488-138491), 61-64(138492-138493, 139017-139018), 117-128(139019-139029, 138460), 139(138461), 141-149(138462-138470), 162-171(138471-138473, 140121-140127), 182-191(140128-140137), 203-212(140138-140139, 141571-141578), 225-234(141579-141588), 243-244(141589-141590), 245n(141591), 246-247(141592-141593), 252-256(141594-141598), 262-264(141599-141601), 444(141602), 450-452(143864-143866), 461(143867), 474-477(143868-143871), 501-504(143872-

VC-5's UH-34D 150211 (c/n 58-1542) displays the Squadron's Safety Award "S" at NAF Naha on 10 January 1973. The camouflage scheme is far less flamboyant than that of VC-5 UH-34E 145716, illustrated earlier. (Photo by Hideki Nagakubo courtesy of the author's collection)

UH-34D 150245 (c/n 58-1622), ship's bird of USS Princeton (LPH-5) during an early-morning preflight inspection. (Photo courtesy of Nick Williams)

Among the Army's helicopter weapons testbeds was H-34A 53-4493 (c/n 58-51). Here, it packs a punch of 24 rockets and 2 .30-caliber machine guns. (Photo PN1281 courtesy of U.S. Army Aviation Museum)

Twin .30 caliber mounts on either side of the fuselage presaged the armed Huey gunships of Vietnam fame. (Photo courtesy of U.S. Army Aviation Museum)

143875), 526-529(143876-143879), 551(143880), 557-559(143881-143883), 562-566(143884-143888), 569-570(143889-143890), 588-592(143891-143895), 596-598(143896-143898), 600(143900), 620-623(143901-143904), 629-634(143905-143910), 646-650(143911-143915), 654-658(143917-143920), 665-669(143921-143925), 676-679(143926-143929), 694-699(143930-143935), 706-709(143936-143939), 722-726(143941-143945), 742-744(143946-143948) 746(143950), 751-754(143951-143954), 759-762(143955-143958), 767-768(143959-143960), 776-780(145660-145664), 820-824(145665-145669), 853(145670), 874(145672), 915(145678)
HSS-1N 899(145674), 906(145675), 909(145676), 912(145677), 923(145679), 926(145680), 929(145681), 934(145682), 943(145683), 947(145684), 951(145685),

955(145686), 962(145687), 966(145688), 970(145689), 974(145690), 982(145691), 987(145692), 992(145693), 997(145694), 1003(145695), 1010(145696), 1014(145697), 1017(145698), 1021(145699), 1024(145700), 1028(145701), 1030(145702), 1038(145703), 1041(145704), 1045(145705), 1047(145706), 1050(145707), 1053(145708), 1055(145709), 1057(145710), 1064(145711), 1069(145712), 1177(147984), 1217-1221(147985-147989), 1230-1236(147990-147996), 1244-1250(147997-148001), 1254-1259(148002-148007), 1266-1272(148008-148014), 1274-1280(148015-148021), 1285-1291(148022-148028), 1297-1301(148029-148032, 148934), 1310-1314(148935-148939), 1324-1327(148940-148943), 1334-1345(148944-148955), 1352-1355(148956-148959), 1363-1366(148960-148963), 1404-1405(149084-149085), 1417-1418(149086-149087), 1430(149841), 1442(149842)

JH-34C 56-4290 (c/n 58-626) underwent tests using a motor-driven feed on a .50-caliber weapon. (Photo courtesy of U.S. Army Aviation Museum)

Choctaws were also used to test various designs for the aerial sowing of mines. (Photo courtesy of U.S. Army Aviation Museum)

Even the Army made use of minimum-space stowage with the Choctaw. CH-34C 54-915 (c/n 58-195) was with the Training School Command at Fort Eustis during April of 1966. (Photo courtesy of Terry Love)

HUS-1 1412-1413(148822, 149318)
UH-34D 1761-1768(152686, 153124-153130), 1777-1779(153133, 153556-153557), 1781-1785(153116-153120), 1792-1804(153697-153704, 154880-154894)
HUS-1A 510(144655), 545-546(144657-144658)
HUS-1L 804(145717), 806(145719), 1240(148112), 1263(148119), 1265(148121), 1273(148122)
LH-34D 1560(150220)

Low-viz schemes were appearing outside of Vietnam by the time CH-34C 54-2887 (c/n 58-287) was photographed in Germany during September of 1968. In the background is 54-2914 (c/n 58-321), with the 351st Aviation Company. (Photo courtesy of Terry Love)

Army

The Navy originally wanted the HSL-1, but contracted for the HSS-1 when the Bell design proved unsatisfactory for the ASW role. The Army (along with the Air Force) clearly preferred the Piasecki H-21 Shawnee's tandem-rotor design over Sikorsky's H-34 transport. Due to Air Force demands on Piasecki's production facilities, the Army could never get enough to meet all of its requirements. Thus, the first of 435

CH-34C 54-931 (c/n 58-222) wears a high-viz Arctic scheme while with the Transport Command Evaluations Group at Fort Eustis in June, 1962. (Photo courtesy of Elof S. Lundh)

H-34A 54-2881 (c/n 58-296) while with the 91st Transportation Battalion, 7th Army, in Germany. (Photo A0571 courtesy of U.S. Army Transportation Museum)

Oklahoma National Guard CH-34C 54-2900 (c/n 58-303) in low-viz medium OD. (Photo N05221 courtesy of MAP)

CH-34C 54-3026 (c/n 373) at Felker Army Airfield in mid-1966. (Photo by Roger Besecker courtesy of the author's collection)

H-34C 54-3034 (c/n 58-389) participates in Operation Short Jab at Aschaffenburg, Germany, in 1961. (Photo PN7427 courtesy of U.S. Army Aviation Museum)

H-34As was delivered from Sikorsky's Bridgeport plant on the last day of 1954.

Operating tests under temperate conditions were conducted at Fort Rucker, Alabama, in July and August of 1955, using aircraft 53-4525 and 53-4528 over a total of 882 flight hours. Shorter tests were conducted in the Mojave Desert using aircraft 53-4517 during July, 1955. Although a variety of recommendations for improvement were made, the Army Aviation Board considered the H-34 satisfactory and recommended it for service. The Air Force, after choosing the H-21 for its own needs, did not concur in this recommendation, and had already advised strongly against Army acquisition of the H-34 in mid-1954. In its report on the *Contract Technical Compliance Inspection of the H-34A Airplane*, the Air Force's Power Plant Laboratory cautioned, "It is believed that the Army, in buying an 'off-the-shelf' item in order to minimize development costs has adopted a policy which will prove to be inadequate."

The first unit to operate the Choctaw was the 506th Transportation Company (Light Helicopter), which began the transition from H-19s in September, 1955. The last was delivered in February of 1959. By its retirement in 1974, the Choctaw had proven its mettle at every level: from Headquarters Flights and small detachments assigned to signals units up through Air Assault Brigades, Transportation Battalions, training commands, and the Executive Flight Detachment. While most served with the Seventh Army in Europe at some point in their history, H-34s were to be found from the Presidio to Italy, and from the Panama Canal Zone to Alaska and Greenland.

With the advent of the all-weather HSS-1N in 1958 and its subsequent success, a program was undertaken to upgrade the H-34A fleet to H-34C standards. This involved the installation of auto-stabilization equipment, making Choctaws night-capable and greatly expanding their mission capabilities. While no Army H-34Cs came from the production line, all existing aircraft had been upgraded by the end of 1964.

One of the myths surrounding the Choctaw tells of the H-34B, with features and capabilities somewhere between the A and C models. Alas, this designation was never applied to Sikorsky's product. Used but seldom noted, on the other hand, were the JH-34A and JH-34C designations, indicating weapons testbeds. While this was a period of great changes in the use of helicopters, it was not a time when the niceties of re-marking aircraft data blocks were always adhered to. There were also three aircraft were designated NCH-34C. Used at the Army Test and Evaluation Command, Fort Huachuca, Arizona, were 54-3016, 54-3032, and 57-1750.

The third Army model was the VH-34A, which consisted of H-34As converted after delivery to VIP standards. The Executive Flight Detachment (EFD), along with VH-34Ds from the Marine Corps' HMX-1, provided helicopters for the Presidential fleet. While both operated in the United States, the EFD aircraft did not travel out of country in support of Presidential missions. A total of eight VH-34As served with the EFD from late 1958 to the end of 1964. All went on to serve with Regular Army and Guard units, and several are now museum pieces.

As with other American H-34s, the Choctaw was instrumental in SAR, disaster relief, and scientific endeavors. In keeping with its original military purpose, Army H-34s saw service in Lebanon during the 1958 military action there. They were ferried from Europe in a series of flights that led to one hundred days of in-country service. Choctaws were also flown from Europe to take part in Joint Task Force Leopold, which operated in the Congo in 1964.

Also like its Navy and Marine Corps counterparts, the Choctaw saw service after the Huey, the Chinook, and the end of the Vietnam War forced it from front-line status. Between August, 1970, and June, 1974, 211 CH-34Cs and VH-34Cs served in the National Guards of twenty-six states, the District of Columbia, and Puerto Rico. When the last CH-34C retired from the Missouri National Guard the same month that

CH-34C 55-4470 (c/n 58-428) with the 3rd Infantry Division, 7th Army, in Germany. (Photo 4523 courtesy of J.M. Friell)

Maine National Guard CH-34C 55-4476 (c/n 58-436) taxis out at Bangor. While there were official color schemes, Guard aircraft do not appear to have been held strictly to them. (Photo Col. 1451 courtesy of Ronald W. Harrison)

H-34C 56-4313 (c/n 58-702) at Le Bourget on 27 May 1961. (Photo 5141 courtesy of J.M. Friell)

the last HH-34J left the Air Force Reserve's inventory, it was the end of American military service for the H-34.

Serials assigned H-34As delivered directly to the Army were: 53-4475 to 4554, 54-882 to 937, 54-2860 to 2914, 54-2995 to 3012, 54-3014 to 3050, 55-4462 to 4504, 56-4284 to 4342, 57-1684 to 1751, 57-1753 to 1771

Construction numbers (prefixed by "58-") and serials (in parenthesis) for H-34As delivered directly to the Army were: 15(53-4475), 21-23(53-4476 to 4478), 28-33(53-4479 to 4484), 39-45(53-4485 to 53-4491), 50-60(53-4492 to 4502), 65-116(53-4503 to 4554, 129-138(54-882 to 891), 150-159(54-892 to 901), 172-181(54-902 to 911), 192-201(54-912 to 921), 213-222(54-922 to 931), 235-242(54-932 to 937, 54-2860 to 2861), 249-251(54-2862 to 2864), 257-261(54-2865

President Eisenhower leaving VH-34C 56-4316 (c/n 58-711)at Camp David. A tail wheel-mounted float is carried in place of the more common belly-mounted float. (Photo 72-3231-15 courtesy of Dwight D. Eisenhower Library)

to 2869), 268-278(54-2870 to 2880), 282-298(54-2882 to 2895, 54-2881, 54-2896 to 2897, 301-309(54-2898 to 2906), 311(54-2907), 313-318(54-2908 to 2913), 321-323(54-2914, 54-2995 to 2996), 325-327(54-2997 to 2999), 332(54-3004), 338-339(54-3000 to 3001), 342-349(54-3002 to 3003, 54-3005 to 3010), 351-352(54-3011 to 3012), 359-362(54-3014 to 3017), 364-369(54-3018 to 3023), 371-373(54-3024 to 3026), 380-386(54-3027 to 3033), 389-394(54-3034 to 3039), 402(54-3040), 404-409(54-3041 to 3046), 411-413(54-3047 to 3049), 415-418(54-3050, 55-4462 to 4464), 423-431(55-4465 to 4473), 434(55-4475), 436(55-4476), 438-443(55-4477 to 4482), 463(5-4485), 471-472(55-4486 to 4487), 480(55-4488), 495(55-4489), 497-499(55-4490 to 4492), 505-506(55-4493 to 4494), 508-509(55-4495 to 4496), 531(55-4497), 540(55-4498), 547-548(55-4499 to 4500), 560(55-4501), 568(55-4502), 574-575(55-4503 to 4504), 616-619(56-4284 to 4287), 624-628(56-4288 to 4292), 642(56-4293), 644-645(56-4294 to 4295, 651-652(56-4296 to 4297), 660-664(56-4298 to 4302), 671-675(56-4303 to 4307), 683-687(56-4308 to 4312), 702-704(56-4313 to 4315), 711(56-4316), 714(56-4317), 716-720(56-4318 to 4322), 728(56-4323), 732-739(56-4324 to 4331), 747(56-4332), 756-758(56-4333 to 4335), 763-765(56-4336 to 4338), 771-774(56-4339 to 4342), 790-794(57-1684 to 1686), 794-796(57-1687 to 1689), 800(57-1690), 814-818(57-1691 to 1695), 825-826(57-1696 to 1697), 842-848(57-1698 to 1704), 865-869(57-1705 to 1709), 877-878(57-1710 to 1711), 884(57-1712), 887-888(57-1713 to 1714), 893-895(57-1715 to 1717), 897(57-1718), 900-905(57-1719 to 1724), 910(57-1725), 918-921(57-1726 to 1729), 924-925(57-1730 to 1731), 927(57-1732), 936-942(57-1733 to 1739), 957-960(57-1740 to 1743), 967-969(57-1744 to 1746), 978-981(57-1747 to 1750), 984(57-1751), 988-989(57-1753 to 1754), 999-1002(57-1755 to 1758), 1008-1009(57-1759 to 1760), 1011-1012(57-1761 to 1762), 1019-1020(57-1763

56-4316 was reassigned to the 3rd Transportation Company at Fort Belvoir after the EFD was disestablished. (Photo 2804 courtesy of AAHS and Herbert C. Lineberger Collections)

to 1764), 1022-1023(57-1765 to 1766), 1025-1027(57-1767 to 1769), 1029(57-1770), 1034(57-1771)

Known VH-34A conversions are 56-4316, 56-4320, 57-1684, 57-1697, 57-1705, 57-1724, 57-1725, 57-1726. All were upgraded to VH-34C standards with the rest of the Army fleet.

The Army's Choctaws were primarily based in Europe and the eastern half of the United States. The Piasecki H-21 Shawnee, on the other hand, was prevalent in the West and Pacific. Thus, it was logical for the H-21 to be the prime Army Aviation helicopter asset in the early years of the war in Vietnam. Some H-34s might have found their way there, however, although the information is shallow. The Army's "2410" database, which provides information on aircraft locations for each month from 1962 forward, shows the following H-34s as having served in Vietnam. Given are the serial and, in parenthesis, dates of service.

93rd Transportation Company: 54-0897 (1-3/63). This was probably an evaluation period, and there are no further entries until the block of aircraft below.

Unit unknown: 53-4492(1-7/70), 53-4507(4-7/70), 53-4509(5-7/70), 53-4515(3-7/70), 53-4523(4-7/70), 53-4533(4-7/70), 53-4536(3-7/70), 53-4538(1-7/70), 53-4541(3-7/70), 53-4544(1-7/70), 53-4549(4-7/70), 54-0883(7/70), 54-2860(2-7/70), 54-2888(1-7/70), 54-2890(3-7/70), 54-2894(3-7/70), 54-2905(2-7/70), 54-2908(1-7/70), 54-2912(2-7/70), 54-2999(3-7/70), 54-3002(3-7/70), 54-3003(2-7/70), 54-3005(4-7/70), 54-3008(6-7/70), 54-3009(3-7/70), 54-3012(6-7/70), 54-3037(7/70), 54-3043(4-7/70), 55-4476(5-7/70), 55-4482(4-7/70), 55-4486(1-7/70), 55-4488(3-7/70), 55-4500(6-7/70), 56-4284(5-7/70), 56-4286(3-7/70), 56-4291(2-7/70), 56-4297(3-7/70), 56-4298(4-7/70), 56-4303(3-7/70), 56-4305(3-7/70), 56-4313(1-7/70), 56-4314(4-7/70), 56-4331(6-7/70), 56-4340(4-7/70), 56-4341(5-7/70), 57-1703(3-7/70), 57-1714(5-7/70), 57-1716(5-7/70), 57-1722(2-7/70), 57-1727(2-7/20). These were all overhauled at the Georgia Depot; after noting service in Vietnam, the Army's database shows all immediately entered National Guard service with various states.

Unit unknown: 57-1732(12/70, 3-9/71).

Unit unknown: 57-1750(6/72); from and to California National Guard.

While this is "firm" evidence of Army H-34s in Vietnam, there is room for doubt as to its accuracy. With one exception (the 93rd TransCo example), there are no unit designations in the database listing. No photos have surfaced showing these aircraft in Vietnam. There is no apparent reason for the Army to have moved H-34s to Vietnam when turbine-powered helos had already been in the field for several years, replacing piston-engined aircraft. Despite the proliferation of unauthorized unit patches even down to the lowest level, there are none to suggest H-34 operations in Vietnam. Finally, the H-34 is not among the types listed for the First Aviation Brigade, which would most likely have been the controlling organization.

A more likely scenario is that they were charged to the VNAF but held in the US for pilot training. Similar arrangements have been known, such as the ownership by Germany of F-104s kept in American markings and used in the US to train Luftwaffe pilots.

Marine Corps

The Marines recognized the need for a large, troop carrying helicopter when it drew up plans in March of 1950 for what would become the Sikorsky HR2S-1/H-37. As an interim step, the Commandant requested HRS-1/H-19s in July of the same year. Development of the HR2S-1 dragged on, and the HRS-1 proved itself less than what the Corps needed despite its laudable record of accomplishments in Korea. The decision

The VIP interior has been replaced with litters on this unidentified VH-34C at Crissy Army Airfield, San Francisco, on 20 October 1967. The patient is being transferred from Travis AFB to Letterman General Hospital following his flight from Vietnam. (Photo PN1888 courtesy of U.S. Army Aviation Museum)

CH-34C 56-4321 (c/n 58-719), Mississippi National Guard, was shot just as the main rotor was engaged. (Photo N01761 courtesy of MAP)

The Idaho National Guard was the last duty station for CH-34C 56-4331 (c/n 58-739). (Photo courtesy of Rollin A. Hatfield)

Ready for engine start in Germany, CH-34C 57-1728 (c/n 58-920) was with the 351st Aviation Company in May, 1968. (Photo courtesy of Terry Love)

Tests in the late 1950s indicated that the engine heating equipment that came with the H-34 was insufficient. An auxiliary heater is typical of the solutions applied, in this case on a Mississippi National Guard CH-34C. (Photo courtesy of Mississippi National Guard)

UH-34J 147887 (c/n 58-527) carries the legend and distinctive emblem of the Naval Test Pilot School at NAS Patuxent River in June of 1970. (Photo courtesy of Terry Love)

was made to pursue a transport version of Sikorsky's ASW HSS-1. In the ensuing battle for appropriations and clear goals, the approval for HUS-1s was not received until 16 June 1955. Like the HRS-1, it was seen as an interim aircraft until the HR2S-1 could be delivered in quantity. This interim machine would serve far longer, and in substantially greater numbers, than the HR2S-1.

HMR(L)-363 received the first HUS-1s, which were flown from the Sikorsky plant in Connecticut to MCAF Santa Ana, California, by squadron crews in early 1957. A total of 515 HUS-1s and UH-34Ds were delivered to the Marines between 8 February 1957 and 19 April 1968. Only nineteen were delivered between 6 January 1964 and the final shipment almost four and a half years later.

Air America's assortment of H-34s was engaged in the "secret war" in Laos before Marine Seahorses arrived in Southeast Asia, and were still there after the last one left. For public consumption of the time, however, the combat story of the UH-34D began with Operation SHUFLY. Approved on 19 March 1962, and implemented on 15 April of that year, SHUFLY brought HMM-362 (and subsequently HMMs-162, -163, -261, -361, -364, and -365) to Vietnam in place of an additional Army H-21 unit. Initially operating out of Soc Trang, in the Mekong Delta region, SHUFLY helicopters soon traded places with an Army H-21 company at Da Nang. The UH-34D was, despite its single-rotor design, a better machine than the Shawnee at altitudes like those of Da Nang.

The political plans for the war at first called for limited American involvement, and for Army and Marine helicopter

HMM-774 UH-34D 144636 (c/n 58-536) at Andrews NAF during October, 1970. (Photo courtesy of Terry Love)

While some UH-34Ds went to the Navy, some UH-34Js traveled the opposite path. No unit is known for 143944 (c/n 58-725). (Photo 3664 courtesy of AAHS Collection)

The famous Bullpup installation on HUS-1 145762 (c/n 58-875). (Photo courtesy of U.S. Army Aviation Museum)

Not seen as often was the 20mm test-mounted on the port side of 145762. Tests were conducted by HMX-1. (Photo courtesy of U.S. Army Aviation Museum)

units to be replaced by Vietnamese Air Force squadrons as rapidly as possible. The reality of the conflict led to SHUFLY's indefinite extension, along with the addition of Marine H-34 squadrons throughout the country. Even the arrival of its replacement, the Boeing CH-46A, on 8 March 1966, did not send the Seahorse out of the combat zone. All CH-46s were ordered grounded on 31 August 1967, after a series of fatal crashes caused by tail pylon failures. It was not until 18 August 1969 that the last six Marine Corps UH-34s in Vietnam were retired from HMM-362 at Phu Bai. Over the course of its seven years in Vietnam, the Seahorse went everywhere and did everything, from its original missions of troop and cargo transport to close air support, medevac, and mosquito abatement. The price for this service was high: 134 UH-34s

were downed over Vietnam for all causes, from a total of 424 Marine helicopters lost in the conflict.

Overshadowed by its service in Vietnam were the UH-34D's other roles around the world. Seahorses were already present in squadron strength in Japan, Okinawa, and Thailand before SHUFLY commenced. HMMs-261, -263, -264, and -361 supported America's quarantine of Cuba during the crisis of October and November, 1962. From April to June of 1965, HMM-263 flew from USS OKINAWA (LPH-3), and HMM-264 from USS BOXER (LPH-4), in support of Marine Corps operations to evacuate and protect Americans in the Dominican Republic. UH-34Ds from HMM-161 were engaged in recovering NASA astronauts and space capsules during Project Mercury. There were crash victim recoveries, SAR

Minus engine and main blades, UH-34D 145759 (c/n 58-871) sits at MCAS Kaneohe in April of 1969. (Photo courtesy of the author's collection)

More common as armament was the M-60 7.62mm machine gun carried in the door of this UH-34D. Even these were, for a time, not allowed by some squadron commanders, despite the lessons learned by the French in Algeria. (Photo courtesy of National Archives)

UH-34D 147148 (c/n 58-1058) with Reserve HMM-766. (Photo courtesy of the author's collection)

Presidential VH-34D 147191 (c/n 58-1142) on 30 April 1972 after restoration at NAS Norfolk. (Photo by Stephen H. Miller courtesy of the author's collection)

missions, and humanitarian flights during flood and famine. Marine H-34s helped develop aerial firefighting techniques, vertical envelopment and replenishment, and helicopter armor and armament. As early as 1958, HUS-1s could be found in several Reserve units, and during the mid-1960s they shared training responsibilities with Navy UH-34Gs and UH-34Js.

In addition to the standard HUS-1, there were nineteen amphibious HUS-1As in the Corps' inventory, all received between June, 1957, and September, 1958. With and without flotation gear, they were part of Marine Aviation in Vietnam. Two have appeared on the US civil register (BuNos 145713 and 145725), along with BuNo 144657 from the Navy's batch of five.

The third Marine Corps version of the Seahorse was the HUS-1Z. Although the Navy and Army modified existing aircraft to VIP standards, all four of HMX-1's HUS-1Z Seahorses came straight from the assembly line in their executive con-

figuration. The original helicopters of "Marine One" were produced over the course of eight months in 1959 and 1960. They served Presidents Eisenhower and Kennedy domestically alongside the Army's VH-34As and VH-34Cs, and accompanied both Presidents on their trips outside the United States. Following the introduction of the HSS-2 into Presidential service, the VH-34Ds carried on as VIP transports within the Corps into the early 1970s. Two (BuNos 147179 and 147201) spent the last of their days with Carson Helicopters in Pennsylvania, not far from where they served the government.

Direct deliveries to the Marine Corps, by type and BuNo, were:
HUS-1 143961-143983, 144630-144654, 145729-145812, 147071, 147147-147160, 147162-147178, 147180-147190, 147192-147200, 148053-148070, 148072-148111, 148113-148118, 148120, 148753-148802, 148804, 148808-148821, 149321-149391, 149395-149402, 150195-150219

Between Presidential service and restoration, 147191 served the 1st Marine Brigade as a VIP transport. (Photo courtesy of author's collection)

President and Mrs. Kennedy after arriving at the White House aboard HUS-1Z 147201 (c/n 58-1611), 16 October 1961. (Photo AR6842-B courtesy of John F. Kennedy Library)

UH-34D 148062 (c/n 58-1175) from HMM-364. (Photo N19697 via MAP)

HMM-363 UH-34D 148077 (c/n 58-1191) in Alaska to support VQ-1 and the Army Security Agency. (Photo courtesy of the author's collection)

UH-34D 150221-150264, 150552-150580, 150717-150728, 153121-153123, 154895-154902
HUS-1A 144656, 144659-144662, 145713-145716, 145718, 145720-145728
HUS-1Z 147161, 147179, 147191, 147201

Direct deliveries to the Marine Corps, by type, construction number (all prefixed by "58-"), and BuNo (in parenthesis), were:
HUS-1 457-460(143961-143964), 464-468(143965-143969), 473(143970), 483-486(143971-143974), 489-493(143975-143979), 496(143980), 514-518(143981-143983, 144630-144631), 522-524(144632-144634), 535-538(144635-144638), 542-544(144639-144641), 552(144642), 571-573(144643-144645), 577-579(144646-144648), 604-608(144649-144653), 612(144654), 712-713(145729-145730), 730-731(145731-145732), 769-770(145733-145734), 784-789(145735-145740), 809-813(145741-145745), 837-841(145746-145750), 849-852(145751-145754), 861(145755), 863-864(145756-145757), 870-873(145758-145761), 875-876(145762-145763), 889-892(145764-145767), 907-908(145768-145769), 911(145770), 913-914(145771-145772), 916-917(145773-145774), 928(145775), 930-931(145776-145777), 933(145778), 935(145779), 948-950(145780-145782), 952-953(145783-145784), 963-965(145785-145787), 972-973(145788-145789), 975-976(145790-145791), 986(145791), 993(145792), 995-996(145793-145794), 998(145795), 1013(145796), 1015-1016(145797-145798), 1018(145799), 1031-1033(145800-145802), 1035-1037(145803-145805), 1039-1040(145806-145807), 1042(145808), 1044(145809), 1048(145810), 1051-1052(145811-145812), 1054(147147), 1058-1062(147148-147152), 1065-1067(147153-147155), 1072-1076(147156-147160), 1078-1088(147162-147172), 1113-1116(147071, 147173-147175), 1132-1134(147176-147178), 1136-1141(147180-147185), 1146-1149(147188, 147186-1147187, 147189), 1151-1152(147190, 147192), 1154-1157(147193-147196), 1159(147197), 1162-1166(147198-147200, 148053-148054), 1168-1176(148055-148063), 1178-1185(148064-148070, 148072), 1187-1216(148073-148102), 1222-1229(148103-148110), 1239(148111), 1243(148113), 1251-1253(148114-148116), 1261-1262(148117-148118), 1264(148120, 1281-1282(148753-148754), 1284(148755), 1292-1293(148756-148757), 1296(148758), 1302-1306(148759-148763), 1315-1323(148764-148772), 1328-1332(148773-148777), 1346-1348(148778-148780), 1350-1351(148781-148782), 1356-1357(148783-148784), 1360-1362(148785-148787), 1367(148788), 1369-1372(148789-148792), 1377-1378(148793-148794), 1380-1387(148795-148802), 1389(148804), 1393(148808), 1396-1403(148809-14816), 1406-1410(14817-148821), 1416(149321), 1419-1428(149322-149331), 431-1441(149332-149342), 1443-1457(149343-149357), 1460-1492(149358-149390), 1494(149391), 1498-1501(149395-149398, 1506-1511(149399-149402, 150195-150196), 1517-1522(150197-150202), 1530-1535(150203-150208), 1540-1542(150209-150211), 1544-1546(150212-150214), 1554-1556(150215-150217), 1558-1559(150218-150219)
UH-34D 1571-1572(150221-150222), 1574(150223), 1579-1581(150224-150226), 1585-1587(150227-150229), 1591-1593(150230-150232), 1598-1601(150223-150236), 1603-1604(150237-150238), 1608-1610(150239-150241), 1616(150242), 1620-1622(150243-150245), 1628-1629(150246-150247), 1633-1635(150248-150250), 1643-1645(150251-150253), 1650-1652(150254-150256), 1659-1661(150257-150259), 1668-1670(150260-150262), 1674-1676(150263-150264, 150552), 1680-1685(150553-150558), 1688-1700(150559-150571), 1703-1713(150572-150580, 150717-150718), 1715-1724(150719-150728), 1786-1788(153121-153123), 1805-1812(154895-154902)
HUS-1A 511(144656), 580-581(144659-144660), 613-

614(144661-144662), 797-799(145713-145715), 803(145716), 8056(145718), 829-831(145720-145722), 858-860(145723-145725), 880(145726), 885(145727, 896(145728)
HUS-1Z 1077(147161), 1131(147179), 1142(147191), 1161(147201)

Coast Guard

Of all the military models of the H-34, the HUS-1G had the smallest production run. Only six were produced, and all went to the United States Coast Guard (USCG). These aircraft were purpose-built, mating HUS-1 airframes with HSS-1N avionics. The main cabin configuration, with a Stokes Basket stretcher rack on the starboard side and radio operator/observer positions to port, was unique to the Coast Guard aircraft. Three were delivered in September, 1959. Two followed in October, and the last was taken on strength in January, 1960. Operations were based at CGAS New Orleans and CGAS St. Petersburg.

The HUS-1G fleet was plagued by unexplained power losses throughout its career, although the powerplant and systems had been in service with the Navy for over a year before the Coast Guard aircraft were delivered. Three losses were directly attributable to those problems. CG 1343 crashed in Tampa Bay on 26 September 1960 while attempting to recover the crew of a downed Air Force B-47. Attempting to recover both the Air Force and Coast Guard crews, CG 1333 crashed in Tampa Bay during the same operation. Both were recovered, but deemed not worthy of repair. They were held at CGAS Elizabeth City for spare parts. On 30 November 1962, CG 1336 was lost at sea forty-three nautical miles from Elmont Key, Florida; again, the cause was an unexplained loss of power under otherwise optimal conditions.

Under the 1962 Tri-Service Designation System, the HUS-1Gs became HH-34Fs. The surviving three (CG 1332, 1334, and 1335) were sent to storage at NAS Norfolk on 24 July 1963. On 16 October of the same year, they were assigned USAF serials (which were used only on paper) and stricken from the Coast Guard inventory. Most reports indicate a MAP sale, and some indicate transfer to Air Force Reserve SAR units. However, all three went directly to Air America in Laos, where c/n 58-1049 was destroyed on 4 January 1964.

CG 1343, recovered from Tampa Bay, eventually went on display at the Florida Military Aircraft Museum in Clearwater.

model	c/n	s/n	notes
HUS-1G	58-985	1332	63-7972, to Air America
HUS-1G	58-1043	1333	crashed 26 Sep 1960
HUS-1G	58-1049	1334	63-7973, to Air America
HUS-1G	58-1056	1335	63-7974, to Air America
HUS-1G	58-1063	1336	crashed 30 Nov 1962
HUS-1G	58-1068	1343	crashed 26 Sep 1960

Air America/Central Intelligence Agency

While it can be argued that Air America (AAM) was not a military organization, the H-34s it operated spent more years in a combat area than those of any regular armed forces in

HUS-1D 148754 (c/n 58-1282) of HMR(L)-262 is supported by HSS-1N 147991 (c/n 58-1231) of HS-7 during the near-tragic conclusion of Gus Grissom's Mercury-Redstone mission. (Photo S61-2819 courtesy of Lyndon B. Johnson Space Center, NASA)

Reserve UH-34D 148783 (c/n 58-1356) about to have its tail pylon unfolded at NAS Alameda in 1969. (Photo U06268 courtesy MAP)

HUS-1 148790 (c/n 58-1370) while the HUS squadrons still were designated HMR(L). (Photo U15905 courtesy of MAP)

the world. AAM H-34 combat operations in Laos predate any in Southeast Asia by either the US Navy or Marines. Support of clandestine American missions in Laos and Thailand began in late 1960, and ended on 3 June 1974. During that time, over one hundred H-34s saw service with Air America. Between 22 January 1961 and 10 May 1972, thirty-eight of them were reported downed over Laos and three over Thailand. Nine of those losses occurred during the first ten months of service. Officially, these were due to the euphemistic "unfriendly action" rather than hostile fire. As Bob Nokes, an AAM H-34 and S-58T pilot, described the situation, "We never came under enemy fire, because there wasn't supposed to be an enemy. A lot of the 'foreign objects' we were damaged by, however, looked like 20mm shells."

The initial Air America inventory came, by Executive Order, from both the Sikorsky production line (eight machines) and Marine Corps stocks in Southeast Asia (twenty-three air-

craft, essentially eliminating an entire combat squadron). Over the years, additional aircraft would come from the Royal Laotian Air Force (RLAF), the South Vietnamese Air Force (VNAF), the Royal Thai Air Force (RTAF), and storage at Davis-Monthan AFB. An unknown number of unofficial, one-mission "loans" were also made by the Marines to AAM for covert operations. Finally, when the S-58 production line was reopened in 1966, ostensibly to provide H-34s for Military Assistance Program sales, thirty-two of the last fifty-eight H-34s went directly to Air America.

Six S-58Ts were registered to Laos, but operated by Air America, beginning in 1971. The conversion to turbine engines was carried out by AAM personnel at Udorn, Thailand. As with other Air America H-34s, these were quickly equipped with armor and camouflage. Three were downed during 1973, and the remaining three were scrapped when AAM operations were concluded.

UH-34D 149386 (c/n 58-1488) of HMM-363 after both aircraft and squadron designations were changed. (Photo U15903 courtesy of MAP)

UH-34D 150237 (c/n 1603) at Niagara Falls on 17 May 1969. (Photo courtesy of the author's collection)

UH-34D 150724 (c/n 58-1720) off Southern California. (Photo courtesy of Rob Mignard)

Flight operations were based at Udorn and the Laotian capital of Vientiane, as well as military centers such as Long Chang in the Royal Laotian Army's Second Military District. As with any combat fleet, these included cargo and personnel movement, medevac flights, and the recovery of downed aircraft. With RLAF H-34s operating primarily in the Vientiane area, the bulk of General Vang Pao's operations on behalf of the Hmong against the Pathet Lao and North Vietnamese Army was supported by Air America helicopters.

As a contract service, AAM provided maintenance for RLAF and *Commission International d'Control* (CIC, International Control Commission) H-34s beginning in 1964. Included in the maintenance conducted at Udorn was the creation of whole airframes from damaged aircraft. The CIC H-34s (which numbered from six to ten, with at least two downed by hostile fire) were supplied by AAM and operated by French crews in support of the 1962 Geneva Accords.

Known Air America H-34s are:

model	c/n	s/n	notes
HSS-1	58-120	H-70	ex-USN 139022
H-34A	58-133		ex-VNAF 54-886; lost May, 1963
HSS-1	58-145	H-69	ex-USN 138466
HUS-1	58-459	H-D	ex-USMC 143963; lost 1961
HUS-1	58-489	H-C	ex-USMC 143975; lost 1961
HUS-1	58-490	H-H	ex-USMC 143976; lost 1961
HUS-1	58-492	H-I	ex-USMC 143978; lost 1961
HUS-1	58-516	H-J	ex-USMC 143983; to VNAF
HUS-1	58-517	H-K	ex-USMC 144630; lost 1961
HSS-1	58-529	H-71	ex-USN 143879; lost 16 Feb 1971
UH-34D	58-537	H-27	ex-USMC 144637; to VNAF
UH-34D	58-538	H-28	ex-USMC 144638
HUS-1	58-542	H-V	ex-USMC 144639; to VNAF
HUS-1	58-552	H-L	ex-USMC 144642; to VNAF 17 Apr 1963
HUS-1	58-571	H-M	ex-USMC 144643; to VNAF 17 Apr 1963
HUS-1	58-572	H-A	ex-USMC 144644; to VNAF

An unidentified UH-34D disperses anti-mosquito chemicals over the Philippine jungle. (Photo courtesy of National Archives)

The last UH-34D delivered to the Marine Corps: 154902 (c/n 58-1812) while with HMM-561. (Photo U01418 courtesy of MAP)

HUS-1G 1332 (c/n 58-985) with a 150-gallon auxiliary fuel tank. The Doppler radar heads are just visible under the fuselage. (Photo courtesy U.S. Coast Guard Historian's Office)

Coast Guard 1333 (c/n 58-1043) during a drill off St. Petersburg. This was one of two HUS-1Gs lost in Tampa Bay during a single rescue. (Photo courtesy of U.S. Coast Guard Historian's Office)

HUS-1	58-578	H-B	ex-USMC 144647; to VNAF
HUS-1	58-579	H-N	ex-USMC 144648; to VNAF 17 Apr 1963
HUS-1	58-608	H-O	ex-USMC 144653; to VNAF 17 Apr 1963
HSS-1	58-646	H-72	ex-USN 143911; lost 1 Oct 1971
HUS-1	58-713	H-P	ex-USMC 145730; lost 1961
HUS-1	58-731	H-Q	ex-USMC 145732; lost 1961
HUS-1	58-752	H-73	ex-USMC 143952; lost 10 May 1972
HUS-1	58-769	H-R	ex-USMC 145733; to VNAF 17 Apr 1963
HUS-1	58-770	H-W	ex-USMC 145734; lost 1961
HUS-1	58-785	H-S	ex-USMC 145736; to VNAF 17 Apr 1963
HUS-1	58-809	H-T	ex-USMC 145741; to VNAF 17 Apr 1963
HUS-1	58-813	H-E	ex-USMC 145745; to VNAF
HUS-1	58-837	H-U	ex-USMC 145746; to VNAF
UH-34D	58-908	H-35	ex-USMC 145769; to RLAF
UH-34D	58-948	H-22	ex-USMC 145780
UH-34D	58-964	H-34	ex-USMC 145786, to RLAF
UH-34D	58-973	H-21	ex-USMC 145789
HUS-1G	58-985	H-17	ex-USCG 1332
UH-34D	58-1018	H-36	ex-USMC 145799; lost 28 Mar 1967

HUS-1G 1334 (c/n 58-1049) would be lost in Laos while with Air America. (Photo courtesy of U.S. Coast Guard Historian's Office)

Coast Guard 1336 (c/n 58-1063) over the Tampa-St. Petersburg area. (Photo courtesy of U.S. Coast Guard Historian's Office)

H-M (c/n 58-571, ex-USMC HUS-1 144643) lifts a Pilatus Porter wing out of a crash site. (Photo courtesy of Air America Archives, History of Aviation Collection, University of Texas at Dallas)

An AAM pilot checks the crash site load on H-U (c/n 58-837, ex-USMC HUS-1 145746). (Photo courtesy of Air America Archives, History of Aviation Collection, University of Texas at Dallas)

UH-34D	58-1040	H-19	ex-USMC 145807; lost 18 Aug 1964
UH-34D	58-1042	H-20	ex-USMC 145808; lost 13 Mar 1965
HUS-1G	58-1049	H-16	ex-USCG 1334; lost 4 Jan 1964
UH-34D	58-1052	H-29	ex-USMC 145812
HUS-1G	58-1056	H-18	ex-USCG 1335
UH-34D	58-1083	H-37	ex-USMC 147167
HUS-1	58-1166	H-F	ex-USMC 148054; lost 1961
HUS-1	58-1170	H-31	ex-USMC 148057; lost 2 Feb 1967
HUS-1	58-1176	H-G	ex-USMC 148063; to VNAF
UH-34D	58-1188	XW-PHC	ex-USMC 148074; lost 31 Jan 1973
UH-34D	58-1203	XW-PHB	ex-USMC 148089; scrapped 31 May 1974
UH-34D	58-1229	XW-PHD	ex-USMC 148110; lost 5 Oct 1973
UH-34D	58-1243		ex-USMC 148113
HSS-1	58-1388	H-X	148803, from Sikorsky; to CIC
HSS-1	58-1390	H-Y/H-48	148805, from Sikorsky; to CIC
HSS-1	58-1391	H-Z	148806, from Sikorsky; to CIC
HSS-1	58-1392	H-11	148807, from Sikorsky; to CIC
UH-34D	58-1393	H-23	ex-USMC 148808; lost 20 Aug 1965
UH-34D	58-1398	XW-PHA	ex-USMC 148811; scrapped 31 May 1974
HUS-1	58-1400	XW-PHE	ex-USMC 148813; scrapped 31 May 1974
HSS-1	58-1415	H-12	149320, from Sikorsky; to RLAF
HUS-1	58-1444	H-32	ex-USMC 149344

Typically, AAM serial H-22 is the only marking carried by ex-USMC UH-34D 145780 (c/n 58-948). (Photo courtesy of Air America Archives, History of Aviation Collection, University of Texas at Dallas)

XW-PHD (c/n 58-1229, ex-USMC UH-34D 148110) was one of several T-noses operated in Laos by Air America. The cockpit has been armored, but the engine area has not. (Photo S58105A courtesy of Sikorsky Aircraft Corporation via Dana Bell)

HUS-1	58-1470	H-33	ex-USMC 149368
HSS-1	58-1495	H-13	149392, from Sikorsky; lost 8 Jul 1964
HSS-1	58-1496	H-14	149393, from Sikorsky; lost 9 May 1966
HSS-1	58-1497	H-15	149394, from Sikorsky
UH-34D	58-1601		ex-USMC 150236
UH-34D	58-1603		ex-USMC 150237; lost 6 Dec 1973
UH-34D	58-1682		ex-USMC 150555; lost 20 Jan 1971
UH-34D	58-1683		ex-USN 150556; lost 18 Apr 1971
UH-34D	58-1708	H-30	ex-USMC 150577
UH-34D	58-1723	XW-PHY	ex-USMC 150727; lost 30 Mar 1973
UH-34D	58-1726	H-24	ex-RTAF 63-13006
UH-34D	58-1727	H-25	ex-RTAF 63-13007
UH-34D	58-1728	H-26	ex-RTAF 63-13008; lost 5 Mar 1965
UH-34D	58-1762	H-42	153124; lost 19 May 1966
UH-34D	58-1763	H-38	153125; lost 3 Aug 1967
UH-34D	58-1764	H-39	153126; lost 3 Sep 1971
UH-34D	58-1765	H-40	153127; lost 21 Jun 1968
UH-34D	58-1766	H-41	153128
UH-34D	58-1768	H-43	153130; lost 8 Aug 1967
UH-34D	58-1777	H-44	153133
UH-34D	58-1778	H-45	153556
UH-34D	58-1779	H-47	153557; lost 14 Sep 1967
UH-34D	58-1780	H-46	153558; lost 9 Mar 1970
UH-34D	58-1789	H-49	153695; lost 19 May 1970
SH-34J	58-1790	H-50	153696; lost 17 Jul 1969
UH-34D	58-1791	H-51	153697; lost 7 Jul 1967
UH-34D	58-1792	H-52	153698; lost 22 Feb 1968
UH-34D	58-1793	H-53	153699
UH-34D	58-1794	H-54	153700
UH-34D	58-1795	H-55	153701; lost 18 Jan 1968
UH-34D	58-1796	H-56	153702
UH-34D	58-1797	H-57	153703
UH-34D	58-1798	H-58	153704
UH-34D	58-1799	H-59	154889
UH-34D	58-1800	H-60	154890
UH-34D	58-1801	H-61	ex-RLAF 154891
UH-34D	58-1802	H-62	154892; lost 19 Oct 1972
UH-34D	58-1803	H-63	154893
UH-34D	58-1804	H-64	154894
UH-34D	58-1814	H-65	156592
UH-34D	58-1815	H-66	156593; lost 16 Jun 1970
UH-34D	58-1816	H-67	156594
UH-34D	58-1817	H-68	156595; lost 27 Sep 1970
UH-34D	58-1818	H-74	156596
UH-34D	58-1819	H-75	156597; lost 9 Mar 1970
UH-34D	58-1820	H-76	156598; lost 10 Apr 1973

H-23 (c/n 58-1393, ex-USMC UH-34D 148808) at the end of a medevac mission. (Photo courtesy of Air America Archives, History of Aviation Collection, University of Texas at Dallas)

Before being camouflaged and armored, XW-PHA (c/n 1398, ex-UH-34D 148811) flew briefly in Air America's "official" colors". (Photo courtesy of Paul Gregoire)

Known CIC H-34s are:

model	c/n	s/n	notes
HUS-1	58-572	CIC-6	ex-VNAF 144644
HUS-1	58-578	CIC-5	ex-VNAF 144647
HSS-1	58-1388	CIC-1	148803; to RLAF
HSS-1	58-1390	CIC-2	148805; to RLAF
HSS-1	58-1391	CIC-3	148806; to RLAF
HSS-1	58-1392	CIC-4	148807; to RLAF

Air Force

While UH-34Ds were assigned Air Force serials and the HH-34D designation to cover transfers under MAP and MDAP, the USAF did not begin to operate H-34s of its own until after production had ceased. Last to take them on strength of the American branches of service, the Air Force was also the last to operate H-34s.

The end of H-30 (c/n 58-1708, ex-USMC UH-34D 150577). (Photo courtesy of Air America Archives, History of Aviation Collection, University of Texas at Dallas)

The Porter's wings will be strapped to the port side to balance the load carried by H-45 (c/n 58-1778, BuNo 153556). This was one of the batch of UH-34Ds delivered straight from Sikorsky to AAM. (Photo courtesy of Air America Archives, History of Aviation Collection, University of Texas at Dallas)

The CIC aircraft were unmistakable in their white scheme, designed to prevent accidental groundfire. They were also obvious targets. UH-34D CIC-6 (c/n 58-572, BuNo 144644) was once AAM H-A. (Photo courtesy of Air America Archives, History of Aviation Collection, University of Texas at Dallas)

On its way to the RLAF, CIC-1 (c/n 58-1388) carries the last-three of its HSS-1 BuNo, 148803. (Photo courtesy of Air America Archives, History of Aviation Collection, University of Texas at Dallas)

Designated HH-34Js, the Air Force examples were ex-US Navy SH-34Js, taken from storage and stripped of their ASW equipment. These served in the Air Force Reserve (AFRES) from June, 1971, to June, 1974. The 301st Aerospace Rescue and Recovery Squadron (ARRS) operated from Homestead AFB, Florida. In addition to SAR work, the 301st performed Presidential missions and provided security services for NASA at Cape Kennedy. The 302nd ARRS was based at Luke AFB, Arizona. The international airport at Portland, Oregon, was home for the 304th ARRS.

The AFRES HH-34Js were transitional aircraft, filling the gap left by the retirement of the Grumman Albatross amphibian and the needs of the war in Vietnam. The squadrons assigned them never carried complements of more than nine. The 301st didn't fly their first mission with one until more than two months after its receipt. At times they were forced to place aircraft in storage because there were no trained air or ground crews available to use them. By the time parts procurement problems, training, and personnel requirements had been sorted out, the HH-34Js were scheduled for replacement by HH-1Hs and HH-3Es, to begin in January, 1974.

HH-34J 45693 (c/n 58-992) at Homestead AFB. (Photo by W.H. Strandberg, Jr., courtesy of the author's collection)

Also with the 301st ARRS at Homestead, HH-34J 47999 (c/n 58-1246) after a rainstorm on 21 June 1973. (Photo courtesy of the author's collection)

HH-34J 48019 (c/n 58-1278) with the 302nd ARRS, Luke AFB, on 11 August 1973. (Photo courtesy of the author's collection)

By June, 1974, all HH-34Js had been send to Davis-Monthan AFB for storage. From there, many eventually went to the civil market or to base museums. Never more than a brief footnote in Air Force history, most inventory and assignment records for the HH-34Js disappeared shortly after they left service. Even Sikorsky's publicity sheet on their helicopters in Air Force service fails to list the Choctaw.

Known Air Force H-34s are:

model	c/n	s/n	notes
HH-34J	58-563	143885	304th; to US civil
HH-34J	58-706	143936	to US civil N62292
HH-34J	58-915	145678	to US civil N9032F
HH-34J	58-992	145693	301st; to US civil N70709
HH-34J	58-1050	145707	to Costa Rica TI-SPI
HH-34J	58-1057	145710	301st; to LA County Sheriff N87716
HH-34J	58-1246	147999	301st; to US civil N46921
HH-34J	58-1266	148008	to US civil N90822
HH-34J	58-1269	148011	to LA County Sheriff N87717
HH-34J	58-1272	148014	to US civil N62293
HH-34J	58-1278	148019	302nd; to US civil N611PD
HH-34J	58-1280	148021	to US civil N51803
HH-34J	58-1286	148023	to US civil
HH-34J	58-1288	148025	
HH-34J	58-1291	148028	to LA County Sheriff N87718
HH-34J	58-1297	148029	
HH-34J	58-1311	148936	to US civil N90768
HH-34J	58-1313	148938	
HH-34J	58-1325	148941	
HH-34J	58-1326	148942	
HH-34J	58-1327	148943	302nd; to Hill AFB Museum
HH-34J	58-1334	148944	to US civil N944HH
HH-34J	58-1338	148948	to US civil N948HH
HH-34J	58-1343	148953	
HH-34J	58-1353	148957	to Costa Rica TI-SPJ
HH-34J	58-1366	148963	301st; to Warner-Robins AFB Museum

HH-34J 48943 (c/n 58-1327), formerly of the 302nd ARRS, in open storage, November, 1979. (Photo by Werner Hartman courtesy of Terry Love)

In October, 1971, Uruguay's *Aviación Naval* (Naval Aviation) obtained two ex-US Navy SH-34Js. These marked a significant increase in capability, the only other helicopters in Uruguayan Navy service at that time being a pair of Bell OH-13Hs. Serials A-061 and A-062 were assigned respectively to BuNos 143934 and 143941. Tragically, the Seabats collided in midair during a public demonstration over a crowded beach in November of the same year. Eight people were killed, and over thirty injured; the aircraft were destroyed.

Replacements were obtained in 1972, in the form of surplus US Army CH-34Cs. A-063 and A-064 were placed in service, with A-065 and A-066 held in reserve for parts. Only A-064 (by now re-numbered A-066) remained in use into the 1990s, equipping the *Escuadrón Apoyo Logístico* (Logistics Support Squadron). The others were withdrawn in 1988. They went to Hi-Lift Helicopters in Florida in mid-1992, where A-066 joined them and was rebuilt for civil use. Fittingly, perhaps, the aircraft chosen by the *Aviación Naval* as replacements for the H-34s were Westland Wessexes.

model	c/n	s/n	notes
CH-34C	58-200	A-063	ex-54-920; scrapped
SH-34J	58-698	A-061	ex-143934; destroyed
SH-34J	58-722	A-062	ex-143941; destroyed
CH-34C	58-887	A-064/066	ex-57-1713; rebuilt for civil sale
CH-34C	58-942	A-065	ex-57-1739; scrapped
CH-34C	58-1026	A-066	ex-57-1768; scrapped

One of the CH-34Cs obtained after the original HSS-1s crashed, A-064 (c/n 58-887, ex-USA 57-1713) appears in the early medium brown camouflage scheme. (Photo courtesy of Eduardo M. Luzardo)

A-064 became A-066, and sported an aluminum and day-glo scheme until retiring to Florida. The uncovered viewing hatch in the cabin floor is obvious in this shot.(Photo courtesy of Eduardo M. Luzardo)

Chapter 26
Vietnam

Throughout much of the 1960s, Vietnamese Air Force H-34s were the backbone of VNAF helicopter operations against the Viet Cong and North Vietnamese Army. Originally replacing aging Sikorsky H-19s, they came from MAP sales as well as Air America and the US Army, Marine Corps, and Navy. First deliveries came at the end of 1960. Initial training was provided by the US Air Force and Marine Corps on behalf of the Military Assistance Command, Vietnam (MACV), at Nha Trang, and operations commenced in 1963. The declared early aim of phasing out US Marine H-34 squadrons in favor of VNAF units by 1 July 1964 did not, however, come to pass as the war expanded far beyond original estimates.

By 1967, the VNAF operated five squadrons of H-34s, with twenty aircraft per squadron. The 211th Helicopter Squadron, formed in January 1963 from the First Helicopter Squadron, flew from Soc Trang and Binh Thuy. The 213th (originally the Second Helicopter Squadron) was at Da Nang. The 215th started at Bien Hoa, then moved to Nha Trang. The 217th was activated on 1 May 1964 at Da Nang, moved to Tan Son Nhut in July of that year, and to Binh Thuy in December, 1965. The 219th ("King Bees") carried out special operations from Da Nang. This period was the zenith for the venerable workhorse in Vietnam; by the end of 1969, conversion to the Huey was largely complete, and only the 219th continued to fly H-34s.

With the end of their service life in Vietnam, the surviving H-34s were passed on to the Royal Thai and Royal Laotian Air Forces, or returned to the US for storage in the Arizona desert. Of 150 H-34s identified as having served with the Viet-

Fully-marked CH-34C 53-4498 (c/n 58-56) is seen with the 213th Sqd. at Da Nang in 1967. (Photo by Neal Schneider courtesy of Dave Hansen)

UH-34G 138464 (c/n 58-143) survived the war to go into storage at Davis-Monthan. The partial main cabin door is of interest. (Photo courtesy of Tim Kerr)

UH-34G 141575 (c/n 58-209) of the 217th hovers over a rain-slicked runway. There is no armor, and the finish is in good condition, suggesting a fairly new arrival. (Photo courtesy of USAF via Jim Mesko)

Minimally-marked ex-USA CH-34C 54-2903 (c/n 58-306) of the 213th. (Photo courtesy of Terry Love)

902 of the 219th ("King Bees") was once USN UH-34J 143902 (c/n 58-621). (Photo courtesy of Rob Mignard)

UH-34D 145795 (c/n 58-998) of the 213th outbound from a field "somewhere in Vietnam," November, 1966. (Photo courtesy of Terry Love)

namese Air Force (with numbers as high as 231 having been reported), twenty-six are known to have gone to other air forces and seventy-one returned to the US.

As with every other military arm that employed the H-34, the VNAF put it to every task available. Cargo and troop movements, medevac flights, and clandestine operations were part of the H-34s' daily life. Even defoliation missions, better captured in the public eye by photos of C-123s spraying chemicals during low passes over the jungle, started with experimental flights by VNAF H-34s. These began on 10 August 1961, with a test mission north of Kontum using a Helicopter Insecticide Dispersal Apparatus, Liquid (HIDAL) system dispersing Dinoxol.

Cambodia

Because so little is known about it, Cambodian use of the H-34 is nothing more than a footnote to Vietnamese operations. The *Aviation Royale Khmère* (ARK, Khmer Air Force) had two H-34s, obtained when their pilots deserted from the Vietnamese Air Force in 1966. These were assigned to the liaison group that operated from the Pochentong air base at Phnom Penh. At least one was originally supplied to the VNAF through the MDAP program. The fate of these aircraft is unknown, but they probably saw limited service and were eventually scrapped due to a lack of parts and trained personnel.

There are no clear identities for the ARK H-34s. One that has been put forward is serial number 63-13258, but the USAF serials assigned to MDAP H-34s did not include that number. An alternative to this, based on a reading of a photocopy of a magazine photograph, is 63-13208 (c/n 58-1758), which did see VNAF service.

OPPOSITE: From left to right on the 211th's ramp at Soc Trang: ex-AAM UH-34D 144639 (c/n 58-542); UH-34D 143983 (c/n 58-516), also ex-AAM; CH-34C 53-4511 (c/n 58-73), ex-USA; and CH-34C 53-4498 before its transfer to the 213th. (Photo courtesy of USAF via Jim Mesko)

Known VNAF H-34s were:

model	c/n	s/n	notes
UH-34G	58-7	137851	ex-USN
UH-34G	58-8	137852	ex-USN; to US civil
UH-34G	58-9	137853	ex-USN; to US civil
UH-34G	58-13	137857	ex-USN; to US civil
UH-34G	58-14	137858	ex-USN
CH-34C	58-15	53-4475	ex-USN; to MASDC
UH-34G	58-27	138482	ex-USN
UH-34G	58-36	138485	ex-USN; to MASDC
CH-34C	58-40	53-4486	ex-USA; to MASDC
UH-34G	58-46	138488	ex-USN
UH-34G	58-49	138491	ex-USN; to MASDC
CH-34C	58-52	53-4494	ex-USA; to US civil
CH-34C	58-54	53-4496	ex-USA; to MASDC
CH-34C	58-56	53-4498	ex-USA; to MASDC
CH-34C	58-57	53-4499	ex-USA; to MASDC
UH-34G	58-61	138492	ex-USN; to US civil

HUS-1 145777 (c/n 58-931) was damaged in a mortar attack on Bin Thuy. Just identifiable in the right background is its sister ship from the 217th, UH-34D 148756 (c/n 58-1292). (Photo courtesy of USAF via Jim Mesko)

The VNAF experimented with a number of camouflage schemes, such as this three-tone UH-34D from the 213th, October, 1966. (Photo courtesy of Terry Love)

Tan applied over the field green or OD base was the most common disruptive pattern. This unidentified UH-34D is from the 213th at Da Nang, January, 1967. (Photo by Neal Schneider courtesy of Dave Hansen)

UH-34G	58-63	139017	ex-USN	UH-34G	58-146	138467	ex-USN; to MASDC
CH-34C	58-73	53-4511	ex-USA; to MASDC	UH-34G	58-148	138469	ex-USN; to MASDC
CH-34C	58-84	53-4522	ex-USA; to MASDC	UH-34G	58-162	138471	ex-USN; to US civil
UH-34G	58-121	139023	ex-USN; to MASDC	UH-34G	58-166	140122	ex-USN; to US civil
UH-34G	58-123	139025	ex-USN; to US civil	UH-34G	58-169	140125	ex-USN; to MASDC
UH-34G	58-124	139026	ex-USN; to US civil	UH-34G	58-170	140126	ex-USN; to MASDC
UH-34G	58-125	139027	ex-USN; to MASDC	UH-34G	58-188	140134	ex-USN
UH-34G	58-126	139028	ex-USN; to MASDC	UH-34G	58-190	140136	ex-USN; to MASDC
UH-34G	58-127	139029	ex-USN	CH-34C	58-201	54-921	ex-USA; to MASDC
UH-34G	58-128	138460	ex-USN; to US civil	UH-34G	58-209	141575	ex-USN; to MASDC
CH-34C	58-129	54-882	ex-USA; to MASDC	CH-34C	58-219	54-928	ex-USA; to MASDC
CH-34C	58-131	54-884	ex-USA; to MASDC	UH-34G	58-226	141580	ex-USN; to US civil
CH-34C	58-133	54-886	to AAM	UH-34G	58-243	141589	ex-USN; to MASDC
CH-34C	58-134	54-887	ex-USA; to MASDC	UH-34G	58-244	141590	ex-USN; to MASDC
CH-34C	58-138	54-891	ex-USA; to MASDC	UH-34G	58-254	141596	ex-USN; to MASDC
UH-34G	58-141	138462	ex-USN; to Indonesia	CH-34C	58-306	54-2903	ex-USA; to MASDC
UH-34G	58-143	138464	ex-USN; to MASDC	CH-34C	58-366	54-3020	ex-USA; to MASDC

Unarmored CH-34C in tan stripes, 213th Squadron, January, 1967. (Photo by Neal Schneider courtesy of Dave Hansen)

UH-34D with two-tone uppersurfaces and light gray belly, 213th Squadron, January, 1967. (Photo by Neal Schneider courtesy of Dave Hansen)

CH-34C	58-394	54-3039	ex-USA; to MASDC	UH-34J	58-695	143931	ex-USN; to MASDC
UH-34D	58-491	143977	ex-USMC; to MASDC	UH-34D	58-769	145733	ex-AAM
CH-34C	58-509	55-4496	ex-USA; to MASDC	UH-34D	58-785	145736	ex-AAM
UH-34D	58-514	143981	ex-USMC	UH-34D	58-809	145741	ex-AAM
UH-34D	58-516	143983	ex-AAM	UH-34D	58-813	145745	ex-AAM
UH-34J	58-526	143876	ex-USN; to MASDC	UH-34D	58-837	145746	ex-AAM
UH-34D	58-537	144637	ex-AAM; to RLAF	UH-34D	58-841	145750	ex-USMC; to US civil
UH-34D	58-542	144639	ex-AAM	UH-34D	58-850	145752	ex-USMC; to US civil
UH-34D	58-552	144642	ex-AAM	UH-34D	58-931	145777	ex-USMC
UH-34J	58-559	143883	ex-USN	UH-34D	58-950	145782	ex-USMC; to Indonesia
UH-34J	58-564	143886	ex-USN; to Indonesia	UH-34D	58-998	145795	ex-USMC
UH-34J	58-566	143888	ex-USN; to MASDC	CH-34C	58-999	57-1755	ex-USA; to MASDC
UH-34D	58-571	144643	ex-AAM	CH-34C	58-1011	57-1761	ex-USA; to MASDC
UH-34D	58-572	144644	ex-AAM; to CIC	UH-34D	58-1013	145796	ex-USMC
UH-34D	58-578	144647	ex-AAM; to CIC	CH-34C	58-1029	57-1770	ex-USA; to MASDC
UH-34D	58-579	144648	ex-AAM	UH-34D	58-1048	145810	ex-USMC; to MASDC
UH-34J	58-591	143894	ex-USN; to MASDC	UH-34D	58-1169	148056	ex-USMC; to MASDC
UH-34D	58-608	144653	ex-AAM	UH-34D	58-1173	148060	ex-USMC; to MASDC, Indonesia
UH-34J	58-620	143901	ex-USN; to US civil				
UH-34J	58-621	143902	ex-USN; to MASDC	UH-34D	58-1176	148063	ex-AAM; to MASDC
UH-34J	58-622	143903	ex-USN; to MASDC	UH-34D	58-1190	148076	ex-USMC; to MASDC
UH-34J	58-623	143904	ex-USN, MASDC; to MASDC	UH-34D	58-1292	148756	ex-USMC; to RTAF
				UH-34D	58-1440	149341	ex-USMC; to RTAF
UH-34J	58-629	143905	ex-USN; to MASDC	UH-34D	58-1479	149377	ex-USMC; to MASDC, Indonesia
UH-34J	58-631	143907	ex-USN, MASDC; to MASDC				
				UH-34D	58-1480	149378	ex-USMC; to RTAF
UH-34J	58-647	143912	ex-USN; to MASDC	UH-34D	58-1615	63-8248	MAP; to RLAF
UH-34J	58-654	143916	ex-USN; to MASDC, Indonesia	UH-34D	58-1629	150247	ex-USMC; to MASDC, Indonesia
UH-34J	58-667	143923	ex-USN	UH-34D	58-1640	63-8249	MAP; to RLAF
UH-34J	58-668	143924	ex-USN; to MASDC	UH-34D	58-1641	63-8250	MAP; to RLAF
UH-34J	58-669	143925	ex-USN; to MASDC	UH-34D	58-1642	63-8251	MAP; to RLAF
UH-34J	58-677	143927	ex-USN; to MASDC	UH-34D	58-1653	63-8252	MAP; to RLAF

UH-34D in tan, OD, and green over medium gray, 213th Squadron, January, 1967. (Photo by Neal Schneider courtesy of Dave Hansen)

UH-34D	58-1654 63-8253	MAP; to RLAF
UH-34D	58-1655 63-8254	MAP; to RLAF
UH-34D	58-1686 63-8255	MAP; to RLAF
UH-34D	58-1687 63-8256	MAP; to RLAF
UH-34D	58-1701 63-8257	MAP; to RLAF
UH-34D	58-1702 63-8258	MAP; to RLAF
UH-34D	58-1714 63-8259	MAP; to RLAF
UH-34D	58-1726 63-13006	MAP; to RLAF
UH-34D	58-1727 63-13007	MAP; to RLAF
UH-34D	58-1728 63-13008	MAP; to RLAF
UH-34D	58-1729 63-13009	MAP; to RLAF
UH-34D	58-1730 63-13010	MAP; to RLAF
UH-34D	58-1731 63-13011	MAP; to RLAF
UH-34D	58-1734 63-13012	MAP; to RLAF
UH-34D	58-1735 63-13013	MAP; to RLAF
UH-34D	58-1736 63-13014	MAP; to RLAF

UH-34D	58-1738 63-13190	MAP
UH-34D	58-1739 63-13191	MAP
UH-34D	58-1740 63-13192	MAP
UH-34D	58-1741 63-13193	MAP
UH-34D	58-1742 63-13194	MAP
UH-34D	58-1743 63-13195	MAP
UH-34D	58-1744 63-13139	MAP; to RLAF
UH-34D	58-1745 63-13196	MAP
UH-34D	58-1746 63-13197	MAP; to MASDC
UH-34D	58-1747 63-13140	MAP; to RLAF
UH-34D	58-1748 63-13198	MAP
UH-34D	58-1749 63-13199	MAP
UH-34D	58-1750 63-13200	MAP
UH-34D	58-1751 63-13201	MAP
UH-34D	58-1752 63-13202	MAP
UH-34D	58-1753 63-13203	MAP; to MASDC
UH-34D	58-1754 63-13204	MAP
UH-34D	58-1755 63-13205	MAP
UH-34D	58-1756 63-13206	MAP; to MASDC
UH-34D	58-1757 63-13207	MAP
UH-34D	58-1758 63-13208	MAP
UH-34D	58-1759 63-13209	MAP
UH-34D	58-1760 63-13210	MAP
UH-34D	58-1788 153123	ex-USMC; to RTAF

Except for the Marine-pattern nose armor, this UH-34D from the 213th could be mistaken for an Israeli machine. (Photo by Neal Schneider courtesy of Dave Hansen)

Chapter 27
Pre- and Post-Service: A Photo Gallery

Only forty-five commercial S-58s rolled off the production line. Most of the aircraft that have seen civil service, then, were built as military airframes. Their continued use after "retirement" is a tribute to the type's worth. They came from all over the world, but most had their roots in American, French, Israeli, and German service. Because striking an aircraft from the register is largely a voluntary matter, much of what appears on civil registers today is history: the aircraft will never fly again, and often no longer exist. A few of the aircraft pictured here are seen before they entered military service. Some are shown while in the limbo of military storage at Davis-Monthan. The rest can be loosely grouped into three categories: those in service, those in museums, and those which have been "seduced and abandoned."

H-34A 53-4526 (c/n 58-88) on display at Fort Rucker, May, 1978 (see Chapter 1). (Photo courtesy of William E. Parrish)

Formerly with the US Navy, and then the VNAF's 217th Helicopter Squadron, UH-34G 140122 (c/n 58-166) is cannibalized at Tulare, CA, in May of 1991. (Photo courtesy of John R. Kerr)

Seen at Decatur, AL, in June, 1985, Solley Construction's S-58 N8043V was built as HUS-1 140135 (c/n 58-189).(Photo courtesy of John R. Kerr)

Sikorsky S-58 demonstrator N740A would go on to serve as IDF/AF 15, and then return to civilian service. (Photo credit AAHS Collection)

S-58ET N54AH (c/n 58-328) serves Aris Helicopters in San Jose, CA, far from its first home with the French Air Force as an H-34A. (Photo 960426-1 courtesy of the author)

Imperial Helicopters' S-58E N2518M (c/n 58-336) returns to Fleming Field during a Honeywell weapons systems program. The aircraft was originally an H-34A built for the French Air Force. (Photo by Bob Nelson courtesy of Robert N. Steinbrunn)

Before serving with Sabena as S-58C OO-SHP, and the LuM as OT-ZKN/B14 (see Chapter 4), N878 (c/n 58-350) flew with New York Airways and, illustrated here, Chicago Helicopter Airways. (Photo courtesy of the author)

Having been an HSS-1, HSS-1N, and UH-34J (see Chapter 24), S-58T C-GUSD (c/n 58-658) dispenses aerial anti-armor, vehicle, and personnel mines during DoD tests at Honeywell's Elk River, MN, ordnance proving grounds on 3 January 1984. (Photo courtesy of Robert N. Steinbrunn)

N45726, seen with Hi-Lift Helicopters in Florida and last reported with St. Louis Helicopter Airways, was once CH-34C 56-4307 (c/n 58-675). (Photo courtesy of John R. Kerr)

California Helicopter owns the rights to convert aircraft such as former IDF/AF S-58B 13 to S-58ET N5594C (c/n 58-692). (Photo courtesy of Hank Lapa)

The Maryland State Police operated N62292 (c/n 58-706, ex-USN UH-34J 143936 and USAF HH-34J 43936) from Martin Field, as seen here in November, 1978. (Photo courtesy of Rob Mignard)

VH-34C 56-4316 (c/n 58-711) in storage at Davis-Monthan (see Chapter 24). (Photo courtesy of Dennis H. Kuykendall)

Among the many CH-34Cs to pass through Davis-Monthan was 56-4326 (c/n 58-734), which retired from the Pennsylvania National Guard in 1972. (Photo courtesy of John R. Kerr)

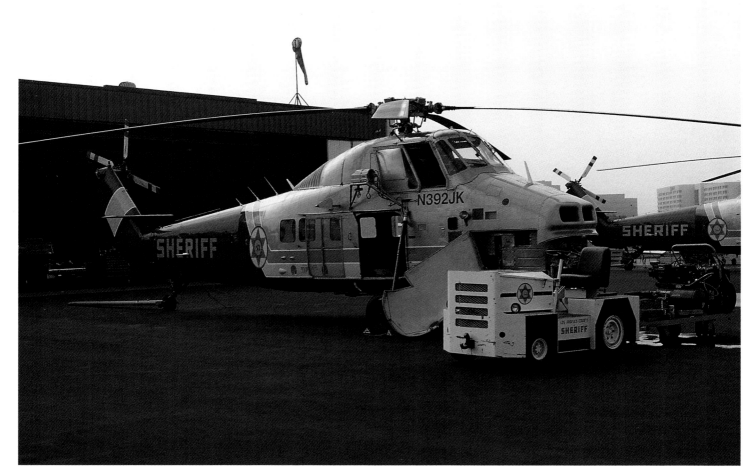

LA County Sheriff's S-58T N392JK (c/n 58-740) was one of the former Argentine Air Force S-58s impressed during the Falklands War (see Chapter 3)

VH-34C 57-1684 (c/n 58-790) survived MASDC to go on display at the Pima County Air Museum in Tucson, AZ. (Photo courtesy of William E. Parrish)

Helcopan of Panama operated HP-612 (c/n 58-802), formerly an H-34G.I of the German *Heerenflieger* and *Luftwaffe*. (Photo C34899 courtesy of MAP)

LH-34D 145717 (c/n 804) on display at the New England Air Museum. (Photo courtesy of Terry Love)

VH-34C 57-1726 (c/n 58-918) retained its tail float while stored at Davis-Monthan. (Photo by H.W. Rued courtesy of the author's collection)

UH-34J 145694 (c/n 58-997) has been lovingly restored for display by the Delaware Valley Historical Aircraft Association at NAS Willow Grove. (Photo by John Benton courtesy of DVHAA)

Coast Guard HUS-1G 1343 (c/n 58-1068) survived Tampa Bay to be displayed in Clearwater, FL. (Photo 950625-17 courtesy of the author)

UH-34D 147171 (c/n 58-1087) on display at NAS Glynco in December of 1986. (Photo courtesy of John R. Kerr)

Winnebago marketed flying campers based on both the S-55 and S-58 airframes, equipped with all the comforts of home including the kitchen sink. One example was N62254 (c/n 58-1193, ex-USMC UH-34D 148079) (Photo courtesy of Winnebago Industries)

LA County Sheriff's S-58DT N724SB (c/n 58-1196) served the Marines as UH-34D 148082. (Photo by Mark Aldrich courtesy of the author's collection)

New Orleans Reserves SH-34J 148000 (c/n 58-1247) partially cocooned at MASDC on 28 January 1979. It would end its days with the destruction of the Heli-Stat. (Photo courtesy of Dennis H. Kuykendall)

SH-34J 148007 (c/n 58-1259) derelict at New Smyrna Beach, Fl. (Photo courtesy of John R. Kerr)

In yet another LA County Sheriff's scheme, H-34J N87717 went on to Heli Flite. It was built as HSS-1N 148011 (c/n 58-1269). (Photo by Bob Neidermeier courtesy of the author's collection)

The remains of USAF HH-34J 48019(c/n 58-1278), registered as N611PD. (Photo 950720-34 courtesy of the author)

It is unlikely that S-58 N51803 (c/n 58-1280, ex-SH-34J 148021) will ever fly again. (Photo 950720-28 courtesy of the author)

UH-34D 148764 (c/n 58-1315) on display at the Greater Southwest Air Museum in July, 1990. (Photo courtesy of John R. Kerr)

S-58D N90561 (c/n 58-1332) of Midwest Helicopters was once UH-34D 148777 (see Chapter 1). (Photo 080080-15 courtesy of the author)

EPA N95958 (c/n 58-1354) operated out of Las Vegas after Navy service as SH-34J 148958. (Photo N15057 courtesy of MAP)

Piasecki's Heli-Stat project was one possible use of retired H-34s, four of which would provide power for a lighter-than-air bag in logging operations. One of the airframes used in the ill-fated attempt was former HT-8 UH-34J 148960 (c/n 58-1363). The extent of the modifications needed for the Heli-Stat are apparent in this photo. (Photo courtesy of Kevin W. Pace)

HH-34J 48963 (c/n 58-1366) in crowded conditions during restoration at the Air Force Museum. (Photo 070482-47 courtesy of the author)

"The Screaming Mimi" appeared in the *Riptide* TV series. S-58DT N698 (c/n 58-1519) went on to fly with Tundra Helicopters. It was built as HUS-1 150199. (Photo courtesy of the author's collection)

Peruvian OB-T-1008 (c/n 58-1547) is seen in Chapter 1 in its later guise as S-58ET N4247V (c/n 58-1547, ex-H-34G.III 150754). (Photo C18523 courtesy of MAP)

The original Thai Ag 912, (c/n 58-1564, ex-H-34G.III 150764), with a UH-60 nose, at Kissimmee, FL, in January, 1994. (Photo courtesy of John R. Kerr)

UH-34D 150227 (c/n 58-1585) at the National Museum of Naval Aviation, NAS Pensacola, before being remarked as an LH-34D. (Photo courtesy of William E. Parrish)

Norwegian LN-OSE (c/n 58-1617) started as H-34G.III 150812. (Photo C15389 courtesy of MAP)

150812 also has appeared on the Spanish register as EC-DDR. (Photo C16058 courtesy of MAP)

Midwest Helicopter's S-58ET N580US (c/n 58-1673) has also been known as German H-34G.III 150804, Norwegian LN-OSF and LN-OSA, British G-BFAG, Canadian C-GAIV, and New Zealand's ZK-HNJ. (Photo 921111-9 courtesy of the author)

Seduced and abandoned: Sud Est-built AdA H-34As SA-175, SA-83, and SA-168 (see Chapter 10) in the Indiana rain during the Summer of 1982. (Photo 070782-16 courtesy of the author)

Appendix A:
Sikorsky S-58 Production by Model and Year

	53	54	55	56	57	58	59	60	61	62
XHSS-1	1	3								
HSS-1		2	57	48	90	51	2			
HSS-1N						23	20	11	93	9
HUS-1					51	71	64	67	77	102
HUS-1A					7	15				
HUS-1G							5	1		
HUS-1L					1	1		5		
HUS-1Z							3	1	1	
H-34A		1	74	210	145	122	27	7		
H-34G.II							20			
H-34G.III										66
S-58			1	4	5		1			
S-58B			4	6		4	1			
S-58C				8	7	2				
S-58D										

	63	64	65	66	67	68	69	model total	
XHSS-1								4	
HSS-1								255	(1)
HSS-1N	16	2		6	3			183	(2)
HUS-1	91	20	2	20	21	6	2	594	
HUS-1A								22	
HUS-1G								6	
HUS-1L								7	
HUS-1Z								5	
H-34A								586	(3)
H-34G.II								20	
H-34G.III	43	21	1					98	
S-58								11	
S-58B								16	
S-58C								17	
S-58D		1						1	

(1) Includes sole HSS-1F turbine testbed
(2) Includes H-34G.IIIM for Germany
(3) Includes H-34G.I for Germany

Appendix B:
S-58 Designations

American, Before and After the Tri-Service System

Air Force

none - CH-34C   used for aircraft transferred under MAP

none - HH-34D   used for aircraft transferred under MAP

none - HH-34J   Search and Rescue (ex-USN UH-34J)

Army (Choctaw)

H-34A - CH-34A	cargo, troop transport, medevac
H-34C - CH-34C	cargo, troop transport, medevac
JH-34A - none	weapons tests
JH-34C - same	weapons tests
none - NCH-34C	electronics tests
VH-34A - none	VIP
VH-34C - same	VIP

Coast Guard

HUS-1G - HH-34F	Search and Rescue

Marine Corps (Seahorse)

HUS-1 - UH-34D	cargo, troop transport, medevac
HUS-1A - UH-34E	amphibious HUS-1/UH-34D
HUS-1L - LH-34D	winterized HUS-1A/UH-34E for Antarctic
HUS-1Z - VH-34D	VIP

Navy (Seabat)

HSS-1 - SH-34G	Anti-submarine Warfare
none - UH-34G	cargo, training
HSS-1N - SH-34J	night-capable Anti-submarine Warfare
none - UH-34J	cargo, training
none - VH-34J	VIP
HSS-1F - SH-34H	one-off turbine testbed

Germany

H-34G.I	equivalent to H-34A
H-34G.II	equivalent to H-34A with auto-stabilization
H-34G.III	SH-34J airframes built to CH-34C standards
H-34G.III Marine	equivalent to SH-34J

Thailand

H.4	all piston-engined models
H.4K	all turbine-engined models

Civil Designations

S-58	cargo (generic Israeli designation for all types used by IDF/IAF)
S-58B	cargo
S-58C	passenger transport (also used by Belgian Air Force)
S-58T/H-34T	turbine-engine conversions of other models by Sikorsky, Orlando Helicopter, and California Helicopter

Other users followed the American series of designations, although the Tri-Service changes were not always put into effect outside of the United States.

Appendix C:
Standard Specifications

Seating capacity: 2 crew (pilot, co-pilot), plus 1 sonar operator, 1 relief (ASW), or 12 to 16 troops, or 8 litters

Engine ratings

    Wright R-1820-84 Cyclone:

        take-off . . . . 1,525 bhp at 2,800 rpm

        normal . . . . 1,275 bhp at 2,500 rpm

    Pratt & Whitney Canada PT6T-3 or PT6T-6 Twin Power Section Turboshaft:

        take-off . . . . 1,625 shp at 33,000 rpm

        maximum continuous . .1,420 shp at 30,360 rpm

Dimensions:

    fuselage length (Cyclone) . . . . . 46 ft. 9 in.

               (turbine) . . . . . . 47 ft. 3 in.

    with pylon folded (Cyclone) . . . . . 37 ft.

               (turbine) . . . . . . 37 ft. 6 in.

    fuselage width . . . . 5 ft. 8 in.

    height overall (tail rotor vertical) 15 ft 11 in.

    main rotor diameter . . . . 56 ft.

    tail rotor diameter . . . . 9 ft. 4 in.

    main landing gear tread (early) . . . 12 ft.

                   (late) . . . . 14 ft.

    cabin interior length . . . . 13 ft. 3 in.

               width . . . . 5 ft. 3 in.

               height . . . 6 ft.

    cabin door height . . . . 4 ft.

               width . . . . 4 ft. 4 in.

Weights:

    gross weight . . . . 13,300 lbs.

    useful load . . . . 5,070 lbs.

    empty weight, with standard equipment . . 8,230 lbs.

Performance

    Cyclone:

        maximum speed at sea level . . . . . 120 mph

        cruising speed . . . . 98 mph

        maximum rate of climb at sea level 1,100 fpm

        service ceiling . . . . 9,500 ft.

        hovering ceiling with ground effect 4,900 ft.

               without ground effect 2,400 ft.

        range (normal fuel plus reserve) . . 225 mi.

    turbine:

        maximum speed at sea level . . . . . 138 mph

        cruising speed . . . . 127 mph

        maximum rate of climb at sea level 1,275 fpm

        service ceiling . . . . 12,000 ft.

        hovering ceiling with ground effect 10,400 ft.

               without ground effect 6,500 ft.

        range (normal fuel plus reserve) . . 299 mi.

Also from the publisher

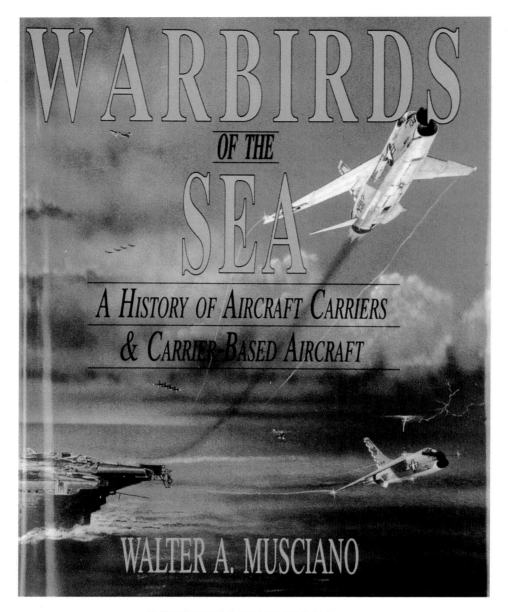

WARBIRDS OF THE SEA:
A HISTORY OF AIRCRAFT CARRIERS
& CARRIER-BASED AIRCRAFT

Walter A. Musciano.

Covers the history and combat career of aircraft carriers and shipboard aircraft from their conception into the future.

Size: 8 1/2" x 11"

over 800 b/w photos, drawings, maps

592 pages, hard cover

ISBN: 0-88740-583-5 $49.95